OUT OF THE FIERY FURNACE

The Impact of Metals on the History of Mankind

'Coalbrookdale by Night,'
by Philip James de
Loutherbourg
(1740–1812), now in the
Science Museum, London.

Robert Raymond

OUT OF THE FIERY FURNACE

The Impact of Metals on the History of Mankind

THE PENNSYLVANIA STATE UNIVERSITY PRESS

University Park and London

Other books by Robert Raymond:

Stirling Moss: A Biography
Black Star in the Wind
Australia's Wildlife Heritage (with Vincent Serventy)
Discover Australia's National Parks
Uranium On Trial (with S. T. Butler and C. Watson-Munro)
The Energy Crisis of 1985 (with C. Watson-Munro)
Australia: The Greatest Island

Published in the United States
of America by The Pennsylvania
State University Press

First published in 1984 by
THE MACMILLAN COMPANY OF AUSTRALIA PTY LTD
107 Moray Street, South Melbourne 3205
6 Clarke Street, Crows Nest 2065

Library of Congress Cataloging-in-Publication Data

Raymond, Robert.
    Out of the fiery furnace.

    Includes bibliography and index.
    1. Technology and civilization. 2. Metallurgy — History.
    3. Civilization — History. I. Title.
CB478.R36   1986   669'.009   86-2367
ISBN 0-271-00440-1
ISBN 0-271-00441-X (pbk.)

Set in 12/13 Garamond
by Savage and Company, Brisbane.
Printed in Hong Kong.

Contents

Introduction

Metals have had such an influence on the growth of civilization, and now play such a pervasive role in modern life, that it might seem presumptuous to consider even outlining the history of metals and metallurgy in a television series, or indeed a single book. I would not have contemplated such an undertaking had I not been given access to the work of Willard C. Lacy, who became my technical adviser and shared a lifetime's study and research with me.

Lacy, a genial but rigorous-minded American, was a practicing mining geologist before becoming Professor of Geological and Mining Engineering at the University of Arizona. After fourteen years at the head of that department, Lacy found himself visiting Australia, and was so attracted by its climate (as well as its geology) that before retiring to Arizona, he took up a post as Professor of Geology at James Cook University in Townsville, Queensland.

While there he carried out consulting work for Australian mining companies and government departments. When I met him he had just completed a worldwide commission to collect material on the history of metals, and to identify the relevant archaeological sites. Lacy did not restrict himself to technology, but explored the profound cultural and sociological impact of metal-using on human society. It was his massive volume of research material, succinctly and lucidly expressed, which made the television series, and this book, possible.

In working on this project I also had a panel of distinguished academic and scientific consultants, whose names are listed at the end of this introduction. With Willard Lacy, they read the film narrations and manuscript of this book, and saved me from many inaccuracies of fact, defects of interpretation, and excesses of opinion. Those that remain are wholly mine.

Some of the consultants, when they read the draft scripts, were concerned at the limitations which the television medium imposes on the telling of history, and in particular its denial of discussion, comparison, qualification, or reflection. Images and words succeed one another in remorseless procession, giving the audience no opportunity to turn back and reconsider a statement, to study an illustration, or even to pause and think. Moreover, by its very nature, the television treatment suggests that events follow one another in a chronological line of development – that things happened first here, and then there. The almost unavoidable conclusion is that someone or some people must have been the 'first' to discover metals, whereupon others followed.

In fact, although the origins of metallurgy are still obscure, we know that it evolved not by a linear progression of individual inspirations, but rather through scattered bursts of innovation and discovery against a broad, slow advance in technology generally. Whether or not an idea took root in any particular area seems to have depended upon whether society was ready to receive it, and economically capable of developing it. Important discoveries about the use of metals were undoubtedly made independently in many parts of the world, but only in some did they lead on to higher levels of technology. (In South America, for example, the Andean peoples discovered how to obtain sufficiently high temperatures with blowpipes to smelt copper but they never developed bellows, and therefore were never able to smelt iron.)

The television series, and this book, tend to emphasize the newer concepts in archaeology and archaeo-metallurgy, which increasingly favor independent discovery and innovation, rather than the established ideas of cultural diffusion. Diffusionists hold that all the most significant advances in human history began in or near the 'cradle of civilization' in the Near East of the Old World, and spread thence across the rest of the planet. In recent years, however, a spate of archaeological and anthropological discoveries has mounted a serious challenge to the diffusionist view, not only of the history of technology but of the origins of such other fundamental human activities as growing food and domesticating animals.

Much of this new thinking is coming from sources outside the traditional centres of study in Europe, and is based on discoveries far from the 'cradle of civilization'. In particular, work by North American, Australian and Asian researchers around the Pacific 'rim' is revealing a vast new panorama of human history.

It is therefore appropriate that *Out of the Fiery Furnace* – underwritten on US public television by an American company, Commonwealth Aluminum – should have been conceived in Australia and made by an Australian film company. In world history, we in Australia have generally occupied a place on the sidelines. But today this is changing, because of our direct involvement in the new sphere of influence which is growing in the Pacific basin, and the part which we might play in the upcoming Pacific Century. It is this role as the more or less innocent bystander which helps us to take an objective view of both history and current events and, in tracing the achievements of man the metalsmith, to give proper credit to the peoples of East as well as West, of the New World as well as of the Old.

In this project I was exceedingly fortunate in being able to resume an old partnership with Michael Charlton, a fellow Australian with whom I had worked in the Australian Broadcasting Commission in the 1960s, and who for the past twenty years has followed a distinguished career in television and radio with the BBC in London, as presenter, narrator, and writer. When he managed to arrange his broadcasting commitments to obtain intervals of free time to work on this series, an enormous problem was solved for me, as both producer and writer. In Charlton I had someone who not only believed in the relevance of the subject, but could make it comprehensible. Because of the exigencies of shooting on five continents over nearly two years, I was often obliged to finalize scripts with Charlton on location. Sometimes he was presented with a concept to digest, improve, and deliver. But there was never a sequence to which he did not contribute a new thought, a vivid phrase, an extra dimension. And as his written and spoken contributions enliven the television episodes, so they permeate this book, and for that I am most grateful.

It is of course unthinkable that such a series could have been made in the time, and to its standard, without the dedication of an outstanding production crew. For practical reasons it had to be small, which made each individual's contributions that much greater. Chris McCullough, my director and associate producer, worked on the project with me from the beginning, and then took on the task of filming a series which was constantly in flux. He brought a sometimes inanimate subject to life, and never failed to find an interesting way of presenting it. Many of his photographs have been used in this book.

Geoff Burton, as director of photography, and David Foreman, his operator, set a visual style for the series which was richly evocative of the widely varied locations and conditions in which they had to work. Ron Lowe and his assistant, Dave Petley, faced with last-minute demands for additional sequences, coped admirably. Leo Sullivan was an indefatigable sound recordist under the most trying circumstances. And John Oakley, a colleague of more than twenty years, not only kept the whole show on the road as production manager, but edited the series in his usual impeccable fashion. The women in all our lives, it goes without saying, deserve medals for their support and encouragement.

In addition to Professor Willard Lacy our consultants were: Professor Don Aitkin, MA PhD, Professor of Political Science, Australian National University; Judy Birmingham, MA, Senior Lecturer in Archaeology, University of Sydney; Professor Geoffrey Blainey, AO, Dean of the Faculty of Arts, University of Melbourne; Christopher Davey, BSc, MA, Senior Lecturer in Mineral Engineering, Royal Melbourne Institute of Technology; Professor Ted Ringwood, FAA FRS, Director of the Research School of Earth Sciences, Australian National University; Dr Howard Worner, CBE DSc DEng (Hon), former Professor of Metallurgy, University of Melbourne.

Robert Raymond

Prologue

The world watched, that July day in 1969, as Neil Armstrong climbed carefully down the ladder of the lunar landing craft and became the first inhabitant of this planet to set foot outside it. One day it may be reckoned the most significant journey that man ever made. But in taking that first tentative step into the future, Armstrong did not forsake the past. One of the first things he did on the Moon was to pick up and study the rocks. It was a reflex as old, perhaps, as man himself.

As our ancestors learned to walk upright and ventured out of the forests on to the plains, they kept their eyes on the ground, even as their horizons widened. The rocks and stones beneath their feet were an endless source of interest and investigation. Eventually, it was the weapons and tools made from the rocks they picked up that enabled the human species to colonise virtually the entire globe. But when their curiosity led them on to extract from the rocks those gleaming substances that we know as metals, they made an advance that changed the landscape of history. That discovery began the long march which over the past ten thousand years has seen the growth of civilisation and the evolution of modern society.

Today, in most parts of the world, the human race has adopted a metal-dependent existence. With few exceptions, every sizeable group of people relies upon metal artefacts to provide the food, shelter, energy, transport and industrial products that make life comfortable. The story of how man reached this stage, of how he discovered and then learned to use metals, is a chronicle of curiosity and imagination, of luck and perseverance, and of uncountable lifetimes in the mines and at the forge and furnace.

The new techniques of exploration and archaeology, such as radio-carbon dating, and the wider view of earth from space provided by satellites, are rolling back history's old horizons and giving us a clearer understanding of the past. Behind the vaulting intellectual accomplishments of mankind in art and music, literature and philosophy, we are beginning to glimpse, through the dust and fumes and smoke of ten thousand years of mining and metal working, the contributions to human comfort and material progress that were made by the begrimed miner and the sweating blacksmith. It is to them that this book is dedicated.

Historic Pathways to Steel

Iron ore

Heated with charcoal up to 850°C

BLOOM IRON
Containing slag

Hammered on anvil

WROUGHT IRON
No carbon

Carburised in charcoal fire

Shaped on anvil

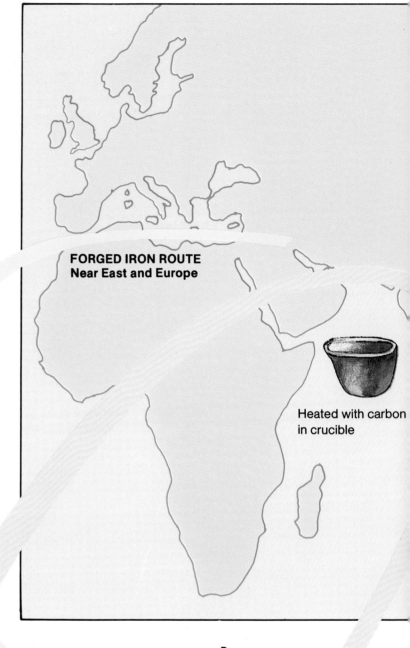

FORGED IRON ROUTE
Near East and Europe

Heated with carbon in crucible

sword

STEELS
Typically between
0.2–1.0% carbon

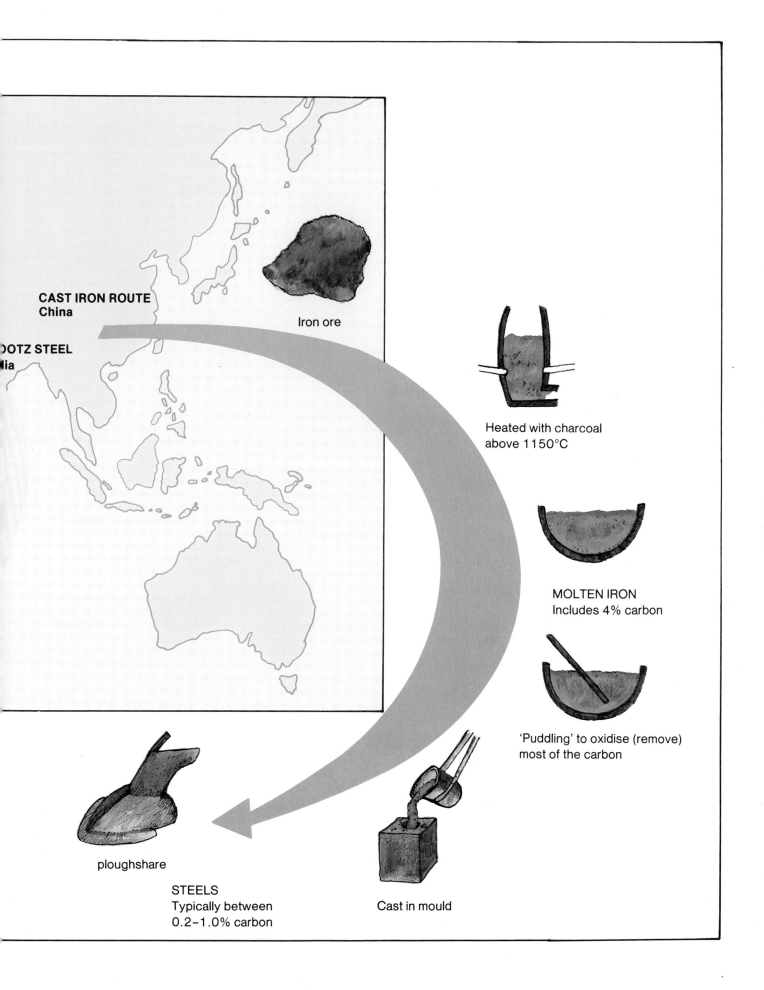

CAST IRON ROUTE
China

ОOTZ STEEL
lia

Iron ore

Heated with charcoal
above 1150°C

MOLTEN IRON
Includes 4% carbon

'Puddling' to oxidise (remove)
most of the carbon

ploughshare

Cast in mould

STEELS
Typically between
0.2–1.0% carbon

Acknowledgements

The author would like to acknowledge the assistance and co-operation of the following individuals, institutions and organisations in the making of the television series and in the preparation of this book:

Abbeydale Industrial Hamlet, Sheffield; Apotheken Museum, Heidelberg; Armco Inc, Butler, Pennsylvania; Beamish Open Air Museum, Durham; Beno Rothenberg, Institute for Archaeo-Metallurgical Studies, London; Bhajan Singh, Ranchi, Bihar, India; British Museum, London; British Steel, London; California State Library, Sacramento; Central Office of Information, London; Charles Eshbach, Michigan Technological University, Houghton; Charles Fries Productions, Hollywood; Chevron Oil Company, Los Angeles; David French, British Institute of Archaeology, Ankara; Department of Antiquities, Athens; Department of Antiquities, Cairo; Department of Antiquities, Cyprus; Department of Antiquities, Damascus; Department of Archaeology, City of Rome; Department of Defence, Washington, DC; Deutsches Museum, Munich; Dr Gerhard Sperl. Leoben, Austria; Drake Well Museum, Titusville, Pennsylvania; Eastman Kodak Company, Rochester, New York; Edinburgh University; Edison Institute, Dearborn, Michigan; Ellesmere Port Boat Museum, Cheshire; Geological and Mining Museum, Sydney; Glasgow University; Glynwed Ironworks, Ironbridge, Shropshire; Golden Spike Historical Site, Utah; Gutenberg Museum, Mainz; Hamersley Iron, Western Australia; Hibbing Taconite, Minnesota; Historical Museum, Peking, HMS "Victory", Portsmouth; Howmet Turbine Corporation, Maryland; Institut Wissenschaftlichen Film, Gottingen, FRG; Institute of Archaeology, London; Ironbridge Gorge Trust, Shropshire; James Mellaart, London University; James D Muhly, University of Pennsylvania, Philadelphia; John Hunt, Union Explosivos Rio Tinto SA, Spain; Kennecott Minerals Corporation, Salt Lake City, Utah; Kew Bridge Engines Trust and Water Supply Museum, London; Kew Gardens, London; Kripal Singh, Jaipur.

Kuringal Gold Museum, Bathurst, NSW; La Rabida Monastery, Huelva, Spain; Laurie Alexander, Australian Embassy, Ankara; Laurie Leskinen, Keweenaw Peninsula, Michigan; Lejre Historical Village, Denmark; Long Senlin, China Film, Peking; Military Museum, Damascus; Monique Bordry, Curie Institute, Paris; Musee de l'Armee, Paris; Museo del Oro, Bogota, Colombia; Museo Oro del Peru, Lima; Museum of Anatolian Civilisations, Ankara; Museum of the University of Pennsylvania, Philadelphia; National Air and Space Administration, Houston, Texas; National Archives and Records Service, Washington, DC; National Coal Board, London; National Museum, Bangkok; National Museum, Copenhagen; National Museum, Jerusalem; Newcomen Preservation Society, Dartmouth, Devon; Nissan, Tokyo; NKK Steel, Tokyo; Orhan Oranli, Konvoy Tur, Ankara; Pechiney, Paris; Pisit Charoenwongsa, Fine Arts Department, Bangkok; Professor Ronald King, Royal Institution, London; Professor R F Tylecote, London University; Rammelsberg Company, Goslar, FRG; Rajvir Singh, Jaipur, and people of Dundlod; Rolls Royce, Bristol; Ruth Meincken, Germany; Saugus Iron Works State Park, Boston; Seiko, Tokyo; Shen Zhihua and staff of the China Film Co-Production Corporpation, Peking; Shin and Mariko Kariya, Shibui Films, Tokyo; Southern California Edison Company; Sovereign Hill, Ballarat, Victoria; SS "Great Britain" Project, Bristol; University of the Arts, Tokyo; Yuendumu Aboriginal community, Central Australia; Zhao Wei, Vice Director, China Film Co-Production Corporation, Peking.

Picture Credits

Pictures are by Robert Raymond, except for those otherwise credited, or acknowledged below: China Film Co-Production Corporation: pp. 69 (bottom), 71. Christopher Davey: p. 31. Christopher McCullough: pp. 2, 66 (left), 100, 104, 114–115, 119, 139, 146, 161, 162, 177, 186, 189, 210, 223 (top), 240–241, 245, 251 (bottom), 257. CRA Exploration Pty Ltd, Canberra: p. 251 (top). Curie Institute, Paris: p. 235. Department of Archaeology, Rome: p 89. Drake Museum, Titusville, Pa.: p. 205. Eastman House, Rochester, NY: p. 248. Edison Institute, Detroit: p. 229. National Archives, Washington, DC: p. 207. Royal Institution, London: p. 227. SS 'Great Britain' Project, Bristol: pp. 179, 181. Smithsonian Institution Travelling Exhibition Service: pp. 39, 40, 41. Angela Raymond p. 4.

CHAPTER 1
From stone to copper

In taking his first steps from the Stone Age into the Age of Metals, man crossed a great divide. But considering his long and skilful use of the other materials around him, that advance took place quite late in human history. In fact in some parts of the world it never took place at all.

In Australia the first arrivals found a continent rich in minerals of all kinds, in places exposed and accessible through aeons of erosion. However, in all their long occupation of this land – which, it now appears, may have begun as early as one hundred thousand years ago – the original Australians made no practical use of those minerals in their metallic form. In a full and self-sufficient existence they were able to satisfy all their needs with what lay easily to hand: stone, wood, bone, clay, fibres, shells, resins, and other natural materials.

Even today the tribes which once roamed the plains have not forgotten their ancestral quarries, where the desert floor is carpeted with chippings from thousands of years of tool making. Nor have they lost the skills to strike the razor-sharp flakes from the flint cores, and to back them with spinifex gum or fix them to shafts. In a single day three old men of the Walbiri people, who now live in the settlement called Yuendumu, north-west of Alice Springs, were able to assemble the tool-kit of knives, choppers, scrapers and spears that enabled the human species to spread across the face of the planet and occupy every habitable continent.

Only in their ceremonial life did the Aboriginal people make any use of the metallic elements which existed all round them. Many of the pigments they used in body decoration and cave painting were made by grinding coloured minerals. From iron ores (which in places were mined) they obtained vivid reds and yellows, from manganese oxide a brownish-black, and from weathered copper ores a range of blues and greens. But in Australia, alone of the inhabited continents, the special properties of the metals themselves went unheeded – until the arrival of the latter-day immigrants.

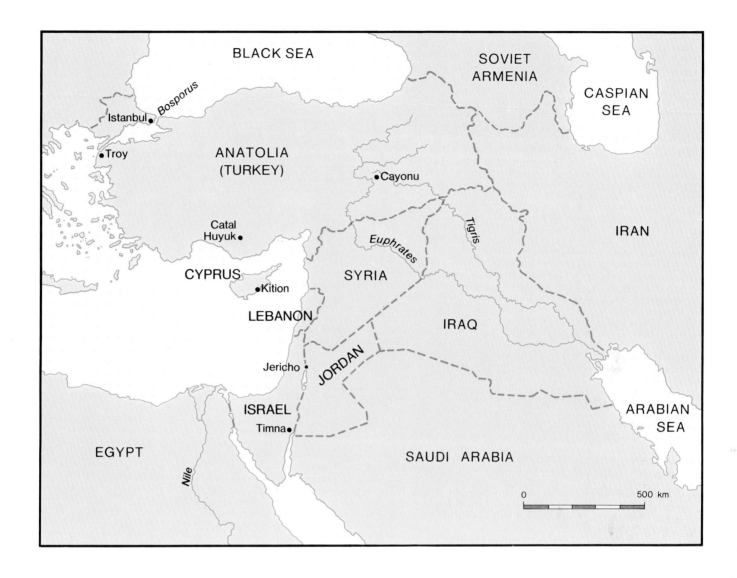

Elsewhere in the world the adoption of metals into everyday life seems to have been associated with the profound changes in human society which began about ten thousand years ago, at the end of the last Ice Age. One of the most significant changes was the transition from the nomadic life of hunting and food gathering to a settled existence in communities. With settlement came the domestication of animals and the cultivation of plants. Thus began what has come to be known as the Neolithic Revolution.

There were now opportunities for people to extend the range of their activities beyond the daily search for food and bare survival from the elements. This period, from about 8000 to 5000 BC, saw the development of the crafts of spinning, weaving and potting. Rapid improvements in domestic architecture followed the invention of the mud brick. There was time and incentive for the diversification of individual occupation, and for the emergence of the specialist. Above all, there was time to think. Ultimately, it might be said, this liberation of the human imagination provided the springboard for almost all progress towards civilisation as we know it today.

Opposite: Thousands of years before metals came into use, a tool-kit derived from stone enabled the human species to colonise every habitable continent. *Elder of the Walbiri tribe of Central Australia at work in a traditional flint quarry, west of Alice Springs.*

3

The appearance of metals in everyday life coincided with the change from the nomadic existence of hunting and food-gathering to settlement in communities, the development of agriculture, and the domestication of animals. *Tribesmen from a wide area of northern India gather for the annual camel sale at Pushkar, in Rajasthan.*

The Neolithic Revolution did not begin in all parts of the world at the same time, because conditions were not everywhere favourable. One region which did favour and encourage it was that part of western Asia called the Near East, and in particular the great belt of hilly country, the 'fertile crescent', extending from Iran along the foothills of the Caucasus mountains to Anatolia and down into the Levant, on the shores of the Mediterranean. This now rather arid zone had a milder and wetter climate at the end of the Ice Age, and the slopes rising from the plains were well forested and alive with game. The eastern end of the crescent, and particularly the region that embraces the upper valleys of the Tigris and Euphrates rivers, has long been recognised as one of the main hubs of the Neolithic Revolution. Recent archaeological discoveries, however, are beginning to delineate an increasingly important role for Anatolia, at the other end of the crescent.

Anatolia forms the western extremity of Asia. It is a large, blunt, mountainous peninsula bounded on three sides by historic seas: on the north by the Black Sea, on the west by the Sea of Marmara, the Dardanelles and the Aegean, and on the south by the Mediterranean. Its eastern frontier abuts Soviet Armenia, Iran, Iraq and Syria. Anatolia comprises more than ninety per cent of the modern state of Turkey,

4

and the capital, Ankara, lies on a windswept plain at its centre. The remaining ten per cent of Turkey lies on the Western side of the Bosporus, the narrow strait linking the Black Sea and the Sea of Marmara. The old capital, Istanbul, known in the past as Byzantium and then Constantinople, bestrides the Bosporus, with most of it on the European side.

This strategic location as the bridge between Europe and Asia has ensured Anatolia a recurring role in human history. It has seen the rise of Troy, and the sudden emergence and equally sudden disappearance of the Hittites; the wealth of Croesus and the touch of Midas; the passage to and fro of the great armies of Xerxes and Alexander; and, in the graceful architecture of Ephesus and Sardis, the first flowering of the Classical period of Aegean civilisation.

But thousands of years before its written history began, this region was the scene of seminal developments whose form and sequence we can discern, at this distance, only by inference. These were the first experiments by Neolithic tribes in the cultivation of the seed-bearing grasses which were eventually to become wheat and barley, and the herding of the wild progenitors of sheep, goats and cattle, which roamed the uplands of Anatolia. And there were other riches in the

One of the earliest places of settlement and metal-working was Anatolia, the mountainous region of Turkey which forms the western boundary of Asia, where it meets Europe at the Bosporus. *The weird sandstone formations of Cappodocia, in eastern Anatolia.*

5

mountains, in places exposed and accessible, whose usefulness to early man we are only now coming to appreciate: copper, lead, gold, silver and mercury.

Given the existence of such resources, it should have come as no surprise when an excavation in the 1960s unexpectedly unearthed in Anatolia the largest Neolithic settlement yet discovered anywhere. This find, however, did reverberate through the archaeological world because Jericho, in Jordan, had always been regarded as representing the peak of Neolithic community development.

The earliest settlement at Jericho, dating from about 8000 BC, was certainly the first known human community to be defended by stone walls. It covered about ten acres, and within the enclosure there were stone watch-towers, and scores of mud-brick houses. At its peak it may have had as many as two thousand inhabitants. But Jericho was overshadowed by the find made in southern Anatolia in 1964 by James Mellaart, who was working at the time at the British Institute of Archaeology in Ankara.

'Anatolia is of course covered with ancient habitation sites inviting excavation,' Mellaart told me in London, where he teaches at the Institute of Archaeology in the University of London, 'but this one had not been previously noted, and presented itself quite luckily. I happened to be travelling across the plateau near Konya when I saw this large, low mound rising out of the plain, surrounded by cultivated fields. It was overgrown with grass and bushes, but when I walked over it I found a few pieces of very early-looking pottery. Nobody in the nearby village of Catal Huyuk, or anyone else I talked to, knew anything about any previous occupation of that mound. So I thought it might be worth a dig.'

Over three seasons Mellaart and his Turkish helpers excavated nearly half a hectare of the mound, which covers in all about fifteen hectares. Then, because of a dispute with the Turkish authorities over another matter, Mellaart had to abandon the work and leave Turkey. It was never resumed. The huge mound at Catal Huyuk remains one of the most tantalising archaeological sites anywhere in the world. But Mellaart had made enough astonishing finds in that one small area to produce an unprecedented picture of Neolithic life, and of a highly developed society in which for the first time both pottery and metals are found in the context of daily existence.[1]

The excavations disclosed twelve successive building levels. Carbon dating indicated an unbroken period of occupation and cultural and economic development of just on a thousand years, from about 6500 BC to about 5650 BC. The closely packed houses were made of sun-dried mud bricks, and were entered through a hole in the roof, reached by a ladder. Most contained a living-room with a wooden sleeping platform, and a storeroom for food. The total population could have been as high as six thousand.

The profusion of animal bones found in the houses and their foundations shows that the people of Catal Huyuk were living in close association with goats, sheep and cattle, while still hunting wild boar, deer, leopard and other game. Agriculture appears to have been well

6

advanced; the storerooms yielded three kinds of wheat grains, two kinds of peas, and barley. Traces of oils extracted from pistachios, almonds and acorns were found in containers.

Technological skills were also developing impressively. Beautifully worked spear points, arrowheads, daggers and polished hand mirrors were made from obsidian, the dark grey natural glass from volcanoes which was highly prized in Neolithic times. In the houses were found beads, amulets and pendants made from coral and coloured minerals, stone vessels for keeping oils and perfumes, and intricate containers of basketwork and wood. From about 6000 BC, black, red and streaked pottery begins to appear. This is among the earliest pottery known anywhere in the world.

Some of the buildings were larger and more elaborate, their walls decorated with hunting scenes in red, black and yellow pigments. There is even a depiction of a settlement with a volcano erupting behind it. This apparently represents Catal Huyuk and the nearby volcano, Hasan Dag, which is now extinct. As such, it must be considered the earliest example of landscape art in human history.

Many of the buildings seem to have been used as shrines. The walls carried large ceremonial heads of bulls and other animals, built up with clay and plaster from natural skulls and horns. These, and small clay figures of a mother goddess giving birth, suggested to Mellaart that what he had excavated was the priestly quarter of a community with the beginnings of a religion based on a fertility cult.

Trade in obsidian in the ancient world established channels for the spread of metal-working technology. *Obsidian, a form of dark glass poured from volcanoes, was greatly prized for the keen edges it provided.*

The particular significance of Catal Huyuk to our story lies in the discovery, among the thousands of artefacts, of a handful of small objects made of copper and lead, together with fragments of mineral ores of copper, lead, iron and mercury. Some of the copper pieces, which were strung like beads on the fringe of a garment, show signs of having been hammered; the cloth has been carbon dated to at least 6000 BC. The only older man-made metal articles ever found are four copper pins or awls from Cayonu, farther east in Anatolia, and a few small copper trinkets from Iran, including a pierced pendant which could possibly date from the ninth millennium BC. But these earlier finds lacked the cultural and economic background of Catal Huyuk, through which we can visualise how metals began to enter daily life in Neolithic times.

Catal Huyuk was clearly an important trading centre, with a comparatively large and wealthy population. Besides a possible agricultural surplus, the people would have had access to a valuable source of income in the obsidian which came from the two nearby volcanoes, both very active at that time. Since no source of obsidian has been found in any of the neighbouring populated areas – Syria, Jordan, Palestine or Cyprus – it is likely that the inhabitants of Catal Huyuk held a monopoly of this desirable material. Finds from the excavation suggest that they traded obsidian all over the Levant, in exchange for such goods as the fine flint of Syria and decorative shells from the Mediterranean coast.[2]

Once the particular properties of metals and their ores began to be recognised they must have become important trade goods, as contacts and regular communications developed between the growing Neolithic settlements of the Near East. And this raises the question of what it was that led early man to begin using metals in the first place. Almost certainly, it was not the compulsion to make better tools or weapons. That imperative would only have produced more improved techniques of working flint and obsidian, in the tradition of millions of years of tool making.

Cyril Stanley Smith, Professor Emeritus at the Massachusetts Institute of Technology and a renowned authority on metallurgical history, has pointed out that 'necessity is not the mother of invention – only of improvement. A man desperately in search of a weapon or food is in no mood for discovery; he can only exploit what is already known to exist. Innovation and discovery require aesthetically-motivated curiosity; they do not arise under pressure of need, although of course once new properties of matter become known they are available for any use.'[3]

All the evidence we have of early metallurgy supports this opinion. Because of the unusual character and initial rarity of metals, they were first used for decoration rather than utility, for ornaments rather than knives. Their wider use had to await the emergence of receptive and economically advanced societies which were able to grasp and exploit the potential of metals.

Catal Huyuk was, apparently, not such a community. Ironically, although it was among the earliest Neolithic centres to handle metals it

8

was unable to benefit greatly from this advantage. A solid copper mace head found at Can Hasan, not far from Catal Huyuk, and dated to about 5000 BC, was certainly a significant advance for this region of the Near East, because it is the earliest indisputable example yet found of metal cast in a mould. But the culture which produced it – and Catal Huyuk – did not survive. Anatolia sank into a stagnation that was to last for two thousand years. Catal Huyuk itself, which in Mellaart's words 'shines like a supernova among the rather dim galaxy of contemporary peasant cultures,' burned itself out. As far as the rise of metallurgy is concerned, Catal Huyuk is an example of a community stumbling upon resources the enormous potential of which it was not yet ready to exploit.

It seems unlikely that we will ever know where or when metals were first used deliberately, but it is possible to speculate about how early man first became aware of metals as a particular class of substances. In all probability it was a gradual process, incidental to his continuing use of other materials. To begin with, it may have been simply the observation that some stones or pebbles behaved differently from others, in that they seemed 'heavier', did not crack or chip when hammered, and in some cases could be beaten into any desired shape.

Such would have been examples of the so-called 'native' metals – that is, metals not combined with other elements in mineral ores but existing in more or less pure state, lying about on the ground or exposed in mineral outcrops. Those most obvious to early man would have been gold, copper and iron; platinum and silver also occur as native metals, but much more rarely.

Gold is the outstanding example of a native metal, because of its natural reluctance to combine with other elements. Gold always exists in a virtually pure state, no matter what its physical form – veins in quartz rock, nuggets on the surface, or fine grains in alluvial sands. It may therefore have been the first metal to catch man's eye and invite his attention. We cannot be sure of this, because no examples of worked gold have survived from Neolithic times. But such a gap in the archaeological record is not surprising. Gold, because of its special properties, has probably always had a high value, and this would ensure that anything made from it would never be discarded, but passed on from generation to generation, until eventually re-worked. Where gold objects were buried in tombs, they would be the first items sought by grave robbers and melted down.

No examples of worked iron have survived from Neolithic times, either, but for a different reason. Unlike gold and other precious metals, iron combines easily with other elements, especially the oxygen in the atmosphere. In the familiar process of rusting, pure iron is rapidly reduced to iron oxide dust. The earliest iron objects we know date back only to the fourth millennium BC, and all of them are very badly corroded. The fact that they have survived as long as they have is due to their surprising origin: the iron they are made from is not native at all, at least as far as this planet is concerned. It came from space, in the form of meteorites.

Meteoric iron contains a high percentage of nickel, which helps the iron to resist oxidation. This nickel content also enables meteoric iron to be clearly distinguished from terrestrial iron, which never contains very much nickel. Therefore it is possible to state quite positively that the earliest iron objects were all made of meteoric iron, picked up from the ground or, as in Greenland, knocked off in slivers from large meteorites weighing many tonnes.

Copper exists in both native and combined states. Where large out-crops of copper ores occur it is not unusual to find pockets of metallic copper among the other minerals. And since copper is much more common than either gold or meteoric iron, it is not surprising that all the earliest known metal artefacts are made of native copper.

Once the special properties of native metals were recognised they would obviously have been sought out and collected, and because they were never particularly abundant it is reasonable to suppose that they must have become increasingly rare and valuable. Across the ancient world, however, the demand for these interesting new materials must inevitably have grown. It was a demand that in the end could only have been met, as it was met, by tapping the vast store of metals locked up in the rocks.

The discovery by Neolithic man of how to smelt metals from their ores, where they are linked by strong chemical bonds to elements such as oxygen, sulphur and carbon, must stand as one of the great achievements in human history. It called for a stretch of the imagination, allied perhaps with intuition, that even now, with the clarity of scientific hindsight, we find difficult to explain.

There were, to begin with, no obvious paths to follow from native metals to smelting. Certainly, men learned very early – possibly as early as the time of Catal Huyuk – that native metals could be softened by heat, and even melted and cast in moulds. But simply heating mineral ores does not release the metal they contain. Furthermore, there would have been little incentive to apply heat to mineral ores, since they bear no resemblance to the metals that they contain. Copper ores are par-ticularly deceptive. The vivid green of malachite – much admired in antiquity for jewellery – gives no hint of the red metal within, com-bined with oxygen and carbon. And yet malachite was almost certainly the first metallic ore to be smelted on a world-wide scale.

The link between native copper and malachite might well have been suggested initially to Neolithic man by the common association of these two forms of the metal in outcrops of ore. But the process by which he then learned how to extract copper from the malachite remains one of the most fascinating questions in all prehistory. It is a major chal-lenge to the new scientific discipline of archaeo-metallurgy, in which archaeologists, prehistorians, anthropologists, geologists and metallur-gists are not only looking at ancient habitation sites and early metal artefacts with new insights, but are using all the tools of modern chem-istry and physics to establish the origins of metal working and the methods of its founders.

To smelt copper from malachite two conditions must be met. One is the application of energy, in the form of intense heat – at least 1084°C, which is the melting point of pure copper. The other is an atmosphere starved of oxygen but rich in carbon. Such an atmosphere draws off the

Copper was the first metal to be widely used, first in its native or naturally pure form, and then through the process of smelting it from its ores. *Above left: The earliest known metal object cast in a mould — a copper mace head from Can Hasan in Anatolia, dated to about 5000 BC, now in the Museum of Anatolian Civilisations, Ankara. Above: Native copper, showing patches of freshly exposed metal and the typical green oxidation which eventually develops on it. Left: Malachite, an oxidised copper ore, was quite common in surface outcrops and was the earliest source of smelted copper.*

oxygen from the heated ore and 'reduces' the copper to a molten metallic form. Given time and sufficient heat, the molten copper separates from the residue of the ore, which forms a lighter molten substance called slag.

This is an exacting process, and it is difficult to imagine it being conceived and applied by Neolithic man other than as the result of some accidental experience, or the observation of some related but haphazard event. Curiosity and intelligent experimentation might then have evolved a method of smelting which worked, even if it was not properly understood.

The circumstances of that accidental experience or haphazard event are not easy to identify, and there is at present no generally accepted explanation. One obvious approach to the problem is to establish where in the ancient world copper ores might have been subjected, by accident or design, to those two criticial conditions: a temperature close to 1200°C, and a 'reducing' atmosphere.

One possibility, which has been strongly supported, is the camp fire. Early man, it is suggested, must on occasion have built his cooking fire

The theory that copper smelting may have evolved from its accidental occurrence in early campfires is now largely discounted, on scientific grounds. *Caravan camp near Jaipur, India.*

on outcrops of ores such as malachite, or perhaps used ore-bearing rocks to make a fireplace. After a particularly hot fire, fanned by the wind, he might have found blobs of smelted copper in the ashes. It is a persuasive idea, and in fact some recent research suggests that it might well have worked for one metal – lead. Experiments at the Institute for Archaeo-Metallurgical Studies, associated with the Institute of Archaeology in London, have shown that lead can be smelted from its common ore, galena, in a wood or charcoal camp fire. Such a fire burns at about 600 or 700°C, and this is well above the melting point of lead, which is only 327°C.[4]

However, where copper is concerned this theory faces difficulties which most authorities on metallurgy find insurmountable. To begin with, the melting point of copper is formidably high – 1084°C – and this is rarely achieved in an open fire, even with wind assistance. A greater problem is the maintenance of a reducing atmosphere. Burning carbon fuel does produce carbon monoxide gas, thus creating a reducing atmosphere immediately above the glowing coals, but in an open fire this is fitful and intermittent. There seems little likelihood of sustaining the two necessary conditions long enough to reduce copper accidentally, or even deliberately, in a camp fire. Most archaeometallurgists are therefore inclined to look elsewhere for the genesis of this critical advance.

One other possibility, which is gaining increasing credibility, derives strong support from two historical facts: both metal smelting and pottery making appeared in Neolithic life at about the same time, and the potter, the first specialist in the management of heat, had under his control all the materials and conditions necessary for the smelting of copper.

The history of pottery making is concerned, like smelting, with the transformation of materials by the application of heat. It no doubt began with the observation that clay is soft and easily shaped while wet, but dries hard in the sun; if wetted again, of course, it softens once more. The natural progression might have been to reinforce or emphasise the drying process in a fire. And so, by about nine or ten thousand years ago, it became known that when clay is fired to about 450°C it undergoes a chemical change and becomes irreversibly hard and waterproof. Above about 1400°C it undergoes a second change: the silica in the clay takes on a glassy structure, and the pottery becomes even more rigid.

At first, pots were heat-treated in open fires. In many parts of the world, including New Guinea, they still are. Later it was found that more predictable results could be obtained by stacking the pots on top of the fuel before lighting it, and covering the pile with earth or other material to keep in the heat and distribute it more evenly. Eventually it was realised that a permanent cover, with a built-in flue, was more efficient than making temporary piles for each firing. Thus, somewhere around the beginning of the sixth millennium BC, the pottery kiln came into existence.

With its thick, heat-retaining walls and flue-assisted natural draught, the pottery kiln could maintain temperatures well in excess of 1000°C for hours. In its enclosed space the fumes from the fire would tend to create a reducing atmosphere. Two of the conditions necessary for the smelting of copper were present. And so, on occasion, was the third requirement: the copper ores themselves.

Very early in the history of pottery, metallic ores had come into use for decoration, perhaps as a direct continuation of the tribal custom of body painting with ground-up mineral ores. The pigments were applied to the pots in liquid form, covered with a glaze – usually made with a lead oxide base – and fired. The pigments took on bright, permanent colours, and the glaze formed a hard, transparent protective coating.

There are many places where pots are still decorated with metallic ores in this way, and fired in kilns whose design has hardly changed in thousands of years. One is Rajasthan, in northern India. The traditional brightly patterned 'blue pottery' produced in and around Jaipur, the capital, is derived from the Islamic style of pottery introduced into India in the sixteenth century. Today there are scores of potters continually engaged, but one who stands out because of his technical grasp of pottery making and the behaviour of metallic pigments is Kripal Singh. His work is in demand all round the world, and is remarkable for its consistency and quality. And yet, as he showed us, Kripal Singh still occasionally produces – quite unintentionally – smelted metal in his kiln.

Kripal Singh uses many different metal ores for pigments, including copper, lead, iron, antimony and cobalt. He makes his glaze by heating lead oxide crystals in the kiln. They melt down into a glassy substance which is ground into a powder, mixed with a binding agent, and applied to the decorated pots before firing. Sometimes Kripal Singh has to throw away a batch of melted-down glaze because it contains globules of pure metallic lead. More rarely, the transformation takes place later, during the firing of the pots, and produces patches of metallic lead in the layer of glaze. (The painted tile from Kripal Singh's kiln illustrated opposite has large splotches of smelted lead obscuring the underlying pattern.)

What is occurring here is clearly the accidental smelting of lead in a pottery kiln. It might not take impossibly remote chance to produce a similar accident with copper ore on the pots. Of course, the rich Islamic style of decoration is not typical of early pottery, which used fewer and simpler patterns, often mere streaks or smears of pigment. Nor could the very early kilns easily maintain temperatures above 1000°C, although finds of a slaggy, greenish over-fired pottery from Ur in Mesopotamia suggest that such temperatures were being reached in kilns by 4000 or 3500 BC.

Despite such reservations, it would be a mistake to underestimate the perceptions or the technological skills of those ancient specialists. As early as 6000 BC, for example, the potters of Catal Huyuk had learned to control the conditions during pottery firing to produce either red or black pots from the same clay. In an oxidising atmosphere – one with air present – the iron oxide in the clay remained red; in a reducing atmosphere – with the air excluded – it turned black. It is not known whether this last effect was produced in a kiln or simply by smothering

Both pottery and metals appear at about the same time in the archaeological record, and it is now thought possible that the use of metallic ores to produce coloured patterns on pots, and the heat and fumes in the kiln, may have combined to create the conditions for the accidental smelting of metals. *Pot from Jaipur, India, decorated with pigments of copper (blue), chromium (green), antimony (yellow), iron (pink), and coated with clear lead glaze.*

The accidental smelting of metal is an observed fact at a pottery kiln in Jaipur, where this decorated tile emerged from firing with the pattern obscured by metallic patches in the normally transparent glaze. *Lead has been reduced from the lead oxide used in the glaze.*

pots on an open fire with earth. But in any case that pragmatic knowledge of reduction would certainly have prepared observant men and women to take advantage, as kilns became more efficient, of any unusual transformation of pigments that might be noticed during pottery firing.

We cannot be sure, of course, that this is how the smelting process was first observed or contrived, but the repetitive nature of pottery firing and the conditions involved make it an obvious possibility. It may be more than coincidence that copper artefacts do not begin to appear in the archaeological record in any quantity until after pottery itself appears. And two areas where high-temperature pottery firing evolved – Mesopotamia and Egypt – went on to develop a high level of copper technology.

If it was in fact the pottery kiln which provided the first clues to smelting, it would soon have been appreciated that such a device was not ideal for the deliberate reduction of metallic ores. In the large air space the reducing gases were not evenly distributed, and much of the heat was wasted. So we can imagine people trying all kinds of methods of creating more effectively the conditions for smelting.

Many directions those early experiments took remain a mystery, but we do know that somewhere along the way the metal workers turned back to, or modified, the open fire. The walls of the fireplace were brought in closer to the fire and raised in height to make a more enclosed space, in which the reducing gases could be concentrated. The copper ore was brought closer to the source of heat by mixing it with the burning charcoal. Air was blown into the heart of the furnace to raise the temperature. And, finally, other substances such as iron ore were added to the copper ore as a 'flux'. Fluxing assists the reduction process, and improves the separation of the molten copper from the slag.

We know this was the way that copper smelting developed, because the earliest indisputable smelting site yet discovered anywhere in the world consists of a bowl furnace which worked exactly as described. It is located in the arid wilderness of the Sinai desert in southern Israel, and it operated in the fourth millennium BC.

Like Anatolia, the Sinai peninsula is one of the crossroads of history. Lying between the Gulf of Suez and the Gulf of Akaba on the Red Sea, it is the meeting point of two continents, Africa and Asia, and of two worlds: the 'heavenly Sinai' of the Exodus and the handing down of the Commandments, of monasteries and pilgrims and holy places, and the 'earthly Sinai' of Egyptian and Roman invasions in search of turquoise and copper, and in later times the wars between Israel and Egypt. The spiritual significance of the Sinai is well documented in the Old Testament and elsewhere, but its role in metallurgical history was not fully realised until the 1960s.

In 1959 an archaeologist, Beno Rothenberg, was exploring the dry wadis of the Timna Valley, near the port of Eilat at the head of the Gulf of Akaba. He had come to know the Sinai as an army commander during Israel's 1956 war; now he was following up earlier archaeo-

logical discoveries of ancient mine workings in the desert, particularly those by Flinders Petrie around the turn of the century.

Among the towering cliffs and wide valleys carved from red, yellow and white sandstone Rothenberg came upon the stone foundations, half-buried by wind-blown sand, of what was apparently a very ancient, walled habitation. Lying about were the unmistakable traces of Neolithic man: flint flakes and shards of pottery. Startlingly, there were also indications of copper smelting: nodules of malachite, and granite mortars and pounding stones for crushing it.

'There were no signs of any kind of furnace,' Rothenberg told me, as we walked around the now exposed wall foundations, 'but I knew they had to be somewhere about. Finally I decided to climb that ridge behind the site. I felt that the ancient smeltermen might have chosen the hilltop to take advantage of any wind, to help raise the heat of their furnace fires.' We climbed the ridge, and on a flat space at the top, littered with stones, Rothenberg described what he had found there.

First he saw a scattering of greenish copper-smelting slag. Some of the pieces contained small 'prills' or blobs of metallic copper. Lying about among the rocks were more flint tools and grinding stones.

The earliest known copper smelting furnace, dated from about 3500 BC, was excavated by Beno Rothenberg at Timna in the southern Sinai in Israel in 1965. *Rothenberg is now Director of the Institute for Archaeo-Metallurgical Studies in London.*

Although he could still find no trace of any furnace, it was enough for Rothenberg to set about organising an excavation.

That first discovery eventually developed into a series of archaeological investigations which continued for more than two decades, interrupted only by the two wars with Egypt, and curtailed finally by the return of the major part of the Sinai to Egypt in 1982. Out of all this has come our most comprehensive understanding of early mining and metallurgy, beginning in Neolithic times and extending through successive Egyptian and Roman occupations of what might now be described as the world's first major 'copper belt'.

When Rothenberg and his team of archaeologists first came back to that hilltop at Timna to begin work, they noticed a large single block of sandstone which looked out of place among the other rocks. It seemed a promising place to start. Digging down in front of the block they gradually uncovered a deep, bowl-shaped depression, its clay sides discoloured by heat. Mixed with the sand which filled it were bits of copper slag and fire-blackened stones. Other pieces of sandstone were found close by, and when fitted together they built up the sides of the furnace to a total depth of about eighty centimetres, and a width at the top of about forty-five centimetres.

Beneath the drifted sand surrounding the furnace they found a working surface, with shards of pottery, flint tools and fragments of charcoal scattered over it. Carbon dating of the charcoal, and the cultural evidence of the artefacts, showed that the site had been used about 3500 BC. What Rothenberg and his team had discovered – and it still stands exposed today on its isolated desert hilltop, within sight of the modern Timna copper mine – was the world's oldest smelting furnace.

When archaeological investigation was extended into the area around the Timna Valley, hundreds more sites connected with mining and smelting were found. Exhaustive studies of the copper prills, ores and smelting slags have been made in laboratories in Israel, Germany, Britain and the United States. Professor R. F. Tylecote, a British historian of metallurgy, has carried out, with others at the Institute for Archaeo-Metallurgical Studies, many smelting experiments with reconstructed furnaces similar to the one found at Timna. From all these studies it is now possible to describe the earliest known method of copper smelting, as it was practised six thousand or more years ago.[5]

The copper ore used at Timna was malachite, which occurs in the area in the form of nodules, embedded in the soft sandstone cliffs. The fuel was charcoal, made from the hard, dense wood of the desert acacia trees. The flux for the smelting was iron oxide, which also occurs in the cliffs.

The choice of fluxing material showed considerable proficiency in metallurgy. The Timna malachite contains a high proportion of silica, in the form of sand particles. During smelting the silica would have formed a viscous, glassy slag, with the copper distributed through it in fine grains. When this cooled and solidified the copper would have been almost impossible to recover. The Timna smeltermen had discovered, however, that iron combined readily with the silica, permitting the copper to form easily recoverable globules in the slag, or sometimes even a small ingot in the bottom of the furnace.

Tests on the slags show that furnace temperatures reached 1180 to 1350°C, which would certainly have required some form of forced draught. No trace of any bellows has yet been found near the original furnace, but it is possible that air was pumped into it with a goatskin bag and a clay tube. The large sandstone block backing the furnace may have served as a support for such an apparatus.

A smelt would last for some hours, with copper ore, flux and charcoal being added from time to time, until a good volume of molten slag had accumulated in the bottom of the furnace. It was then left to cool overnight. Next day the solidified mass of slag and copper was lifted out and crushed with hammer stones to extract the prills of metal. These were re-melted in a clay crucible, so that impurities could be skimmed off. The result was an ingot of almost pure copper.

One use to which the copper was being put in the fourth millennium BC, which also exemplifies the developing skills of the metal workers, is on display at the National Museum in Jerusalem. Following the discovery of the Dead Sea Scrolls, Israeli archaeologists had begun a systematic investigation of the cliffs facing the Dead Sea, and the scores of caves which they contained. In 1963, in one virtually inaccessible cave high up a rock wall, they found a collection of strange metal objects.

In 1963, in an almost inaccessible cave near the Dead Sea in Israel, this hoard of strange metal objects was found, loosely wrapped in cloth. *The cloth has been carbon-dated to about 3000 BC.*

They were loosely wrapped in cloth and wedged into a crevice at the back of the cave, as if hastily hidden. Undisturbed sediments in the cave yielded evidence of human usage dating back into the fourth millennium BC, and the cloth in which the objects were wrapped has been firmly carbon dated to at least 3200 BC.[6]

The collection from the Cave of the Treasures, as it is now called, contains more than two hundred objects of smoothly shaped and polished copper, exuding an aura of both authority and mystery. A few are recognisable as chisels or axe heads. Others appear to be mace heads or crests for mounting on staves, and have cast-in sockets. Many of the objects, however, have no apparent function. They are decorated with the heads of birds or horned animals, including the desert oryx. Some have the form of a coronet, but are too small to be worn.

These are presumably ritual objects, but there is no clue as to who might have made them, or where. Nothing quite like them is known from anywhere else in the ancient world. But the assurance of their execution testifies to a growing mastery of metals and metal working, at a time when stone was still in general use for most purposes, although slowly giving way to copper. This period has been called the Chalcolithic, or Copper–Stone age.

Contrary to what once appeared likely, the transition from the Neolithic to the Chalcolithic no longer seems to have begun in one particular area and then diffused outwards to other lands or continents. In some regions there was a clear borrowing of technology from neighbours, perhaps facilitated by trading contacts, but the accumulating archaeo-metallurgical evidence suggests that copper smelting was discovered independently in many different parts of the world, as human society in those areas reached the level of cultural and economic development necessary to support such an industry.

Certainly, the inhabitants of the 'fertile crescent' were among the first to use the new technology. As Professor R. F. Tylecote explains the growth of metallurgy: 'In the central area surrounding Anatolia the diffusion of ideas amongst the developing cultures of that region was fairly rapid . . . Naturally, the more advanced civilisations, such as Egypt, would be among the first exploiters of the new metallurgical techniques.' And so, at the time when the furnace at Timna was working, articles made of smelted copper were also beginning to circulate in Egypt, Mesopotamia, Syria and Anatolia. These, according to Professor Tylecote, can be distinguished in the archaeological record from artefacts made from native copper by analysis of their chemical and physical structure. One useful guide is the percentage of arsenic or nickel; these are never very high in native copper.[7]

It now appears that copper smelting was not introduced into Europe from Anatolia but began there independently, and perhaps as early as anywhere in the Near East. Recent excavations at Rudna Glava in Yugoslavia have shown that a large underground copper mine was in operation there before 4000 BC. The miners used stone hammers and picks made from deer antlers to dig out the rich veins of malachite. They sank shafts as deep as twenty metres, then drove side tunnels. They cracked the hard rock by building fires against the tunnel face and dousing it with water. Much of their output of copper may have gone to the Vinca people, scattered through the Balkans, who at that time were producing large quantities of copper tools, weapons and ornaments.

Similar Chalcolithic copper mines have recently been identified in southern Spain, some actually among the vast workings of the Rio Tinto mine near Huelva, and others in the nearby limestone hills. Bulgaria, too, is yielding evidence of metal mining and working of hitherto unsuspected antiquity. Graves containing scores of copper objects have been excavated at Varna, on the Black Sea, dated to about 4300 BC. Even more remarkably, they also contained beautiful articles of beaten gold. These objects, which have apparently escaped the attention of grave robbers for six thousand years, may be the earliest examples of worked gold yet discovered.

Present archaeological evidence shows that smelted copper first appeared in Italy between 3000 and 2500 BC, in northern Europe around 1950 BC, in Britain by 1900 BC, and in Scandinavia by 1500 BC. This timetable does suggest a diffusion of knowledge across Europe. A similar staggering of dates eastwards from Iran is indicated by the appearance of smelted copper in India by about 3000 BC, and across the Caucasus in southern Russia by 2000 BC.

However, the date of origin of copper smelting in the Far East, particularly in China and South-east Asia, is still a mystery. In fact it has become one of the most controversial issues in archaeo-metallurgy. Until quite recently, the evidence seemed conclusive that knowledge of smelting and metal working had reached Asia from the West, by rather leisurely paths through India or perhaps central Asia. Now there are strong claims for the independent discovery of metal working in the East, at an astonishingly early date. This new evidence is discussed in the next chapter.

Africa south of the Sahara is another region where assumptions are being challenged. It had long seemed self-evident that Chalcolithic skills in copper smelting and metal working had never spread south or south-west from Egypt. The peoples of Black Africa appeared to have entered their metal-working phase only with iron, in the first few centuries BC, presumably following trade contacts with the Arabs across the Sahara. The famous bronze-casting tradition of Benin and Ife was thought to have begun around the fourteenth century AD, and was undoubtedly derived from north Africa.

But a discovery in 1960 at Igbo Ukwu in south-eastern Nigeria has raised questions about that supposedly late start. An excavation on behalf of the Nigerian Department of Antiquities unearthed the earliest castings so far known from that country. They are mostly figures of a ceremonial nature, richly decorated with stylised animals, and are made of copper, some of it relatively pure, some much less refined. They have been carbon dated to about the ninth or tenth century AD.

One interesting thing about these objects is that they appear to be made of quite primitively smelted ores, unlike the copper prepared north of the Sahara. Some also contain a considerable amount of silver, which would almost certainly have been recovered by Arab or European silversmiths of that period. Everything about them suggests that they were smelted locally. This fact, together with their age, and the recent dating of copper mines in West Africa to the first millennium BC, must at least raise the possibility that Black Africa did have an early copper metallurgy, and may even have discovered copper smelting independently.

Discoveries such as this, and others even more unexpected, are making it increasingly hazardous to maintain negative assumptions about early technology in areas where comparatively little scientific exploration has been carried out. Confirmation of this was provided by the publication in 1982 of the first evidence of prehistoric copper smelting in the New World.

It has long been known that the Indians in South America used native gold, silver and even platinum as early as the first millennium BC, but objects made from copper do not appear until about AD 500. Knowledge of copper working then seems to have spread through Central America into Mexico in pre-Columbian times, although evidence for copper smelting was missing, despite the abundance of copper and other metals, and the obvious competence of the Indian metalsmiths. When it finally came, however, the evidence was overwhelming.

Three Americans, two archaeologists and a geographer, announced the discovery of a large-scale copper-smelting industry at Batan Grande, in the Andes in northern Peru. They described how they had excavated more than twenty bowl furnaces, and estimated that another hundred exist on the site. These are arranged in groups of three or four, about a metre apart, on terraces facing the prevailing wind. This layout was apparently designed to increase the natural draught into the furnaces, and to carry the smelting fumes away from the working area in front of the terraces. The whole site is littered with slag, remnants of clay pipes for blowing the furnaces, and rocking stones and slabs for crushing the slag and extracting the prills of copper.[8]

The level of technology at Batan Grande is about the same as at Timna in the fourth millennium BC, but it did not begin until much later, perhaps AD 800. Smelting appears to have continued for at least five or six hundred years, until well into Inca times. All the copper prills so far studied contain other metals, and it is hoped that trace analysis will help to establish a pattern of distribution of the Batan Grande copper to other parts of Peru.

In North America an entirely different development had taken place, much earlier than in South America. Perhaps as far back as 3000 BC an elaborate copper culture began to grow up round the Great Lakes. It was based on an abundance of native copper, produced by an extraordinary geological accident. With such resources the North American

Because of its attractive appearance and comparative ease of working, copper has remained in demand throughout the history of metals. *Copper market in Damascus, Syria.*

Indians presumably had no need to go on to smelting, because no evidence of it has ever been found.

Thus it can be seen how and why copper came to be the first metal that man used widely, and on a large scale. Copper deposits were to be found on or near the surface in many parts of the world. The smelting process, once grasped, was not particularly difficult to carry out, and good quality copper could be produced with reliability. The metal itself was attractive in appearance and was easy to work. As a result it steadily replaced stone and other natural materials for all kinds of utensils, ornaments and small tools. But when it came to large tools and weapons the very softness of copper was a handicap. It could be hardened to some degree by repeated hammering and re-heating – the technique of annealing – but too much working made it brittle.

There was need for a new advance, and a new metal. That new metal, when it came, was the creation of man the metalsmith rather than the gift of nature. It gave its name to the first great span in the history of civilisation: the Age of Bronze.

CHAPTER 2
The rise of bronze

Finds from various parts of the Near East suggest that towards the end of the fourth millennium BC some metal workers were beginning to manipulate the purity of their copper during smelting to improve its utility.

From the beginning, smelted copper had included traces of other metals, such as silver, lead, arsenic, antimony, iron, and sometimes tin. These had been present in the original copper ores, and were more difficult to remove from the copper during smelting than other elements such as sulphur or silica, which were driven off as fumes or ended up as slag. The proportion of trace metals varied according to the origin of the ores. It would have been natural enough for the smelting specialists in Chalcolithic communities to learn to associate certain identifiable ores with the kind of copper they produced, and perhaps deduce the presence of other metals which gave the copper particular properties.

The next step was a major advance in the production of metals: the smeltermen worked out not only how to remove undesirable elements but, more importantly, how to increase the proportion of those which made copper more useful. The first metal to appear in smelted copper in proportions significant enough to indicate intention was arsenic. Early objects from many parts of the ancient world contain up to seven per cent arsenic. This is more than is usually found in copper ores, and suggests the addition of enriched arsenical ores during smelting.

Arsenical copper is little different from pure copper in smelting or casting behaviour, but it has one great advantage which would have become apparent to Chalcolithic metalsmiths: when worked by hammering it becomes much harder than pure copper. The addition of antimony gives a similar advantage, but for some reason arsenic was much more widely used in ancient times, despite the poisonous fumes it gives off when heated.

In another direction, the metal smelters worked on improving the

casting characteristics of copper. Pure copper was difficult to cast not only because of the need to keep the temperature of the metal above its high melting point, 1084°C, but because it had a tendency to oxidise, and to form bubbles from trapped gases. However, it was undoubtedly noticed that the presence of some other metals in the molten copper made it flow more easily, and over a wider range of temperatures.

Eventually, one metal was identified as the outstanding partner for copper. In its presence the temperature required to melt the copper was substantially reduced, to about 950°C. The molten mixture of the two metals flowed much more freely into moulds, and did not form bubbles. And the cast metal was much harder than pure copper, even before work-hardening. This remarkable ingredient was tin, and the ideal proportion of it was about ten per cent. The alloy of copper and tin came to be known as bronze.

The first true tin bronze – in which the percentage of tin is high enough to suggest that it was a deliberate addition – appears in the archaeological record around 3000 BC. The earliest examples known are from the city states of Mesopotamia, including Ur, but over the next few centuries bronze objects begin to turn up all over the Near East: in Egypt, Iran, Syria, Anatolia and Cyprus.

The invention of bronze, with its greater ease of casting, led to a rapid increase in the output of cast metal artefacts. Intricate and elaborate pieces could now be produced by the 'lost-wax' or *cire perdue* method. The increased consumption of bronze naturally led to more active mining and trading of copper. Some of the most explicit evidence that we have of this has been revealed by the excavations carried out by Beno Rothenberg's teams in the Sinai, in the area around the Chalcolithic smelting furnace at Timna. These have established the existence of a very large-scale copper mining and smelting industry between about 1300 and 1150 BC, under the Egyptian empire of the nineteenth and twentieth dynasties.[9]

This was a time of great expansion of Egyptian power under the Pharaohs, and especially by the Ramses dynasty. Their influence extended right through the Levant to Syria, but the occupation of Palestine and the Sinai was of particular importance to them because of the copper mines there. The Egyptians did not work the mines and smelters around Timna themselves, but organised a large labour force of the local tribes of Midianites. At least eight large centres of mining have been found along a ten-kilometre stretch of the sandstone cliffs. The shafts and tunnels were driven with bronze tools; the pick marks are still sharp and clear on the walls. In some places the tunnels broke into the earlier workings of the Chalcolithic miners, and here the furrows of the bronze picks and the shallow indentations of the stone hammers are mingled across two thousand years of time, between the fourth and the second millenniums BC.

The smelting centres were like modern industrial estates. One of the largest that has been found covered two hectares. It was surrounded by a stone wall, and included storage pits for malachite, charcoal and iron

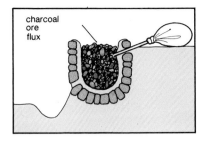

The Egyptian copper smelting furnace was filled with a mixture of copper ore, charcoal and iron ore to act as a flux. It was blown for several hours by foot or hand bellows.

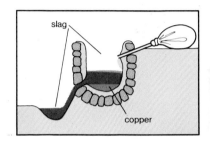

By the end of the smelt the copper had separated from the slag, which was tapped off.

The Timna Valley in the southern Sinai desert was the world's earliest major 'copper belt'. It has been the scene of mining and smelting for at least five thousand years, first by Neolithic tribes, then by the Egyptians and Romans, and now by the Israelis. *The large enclosed area is an Egyptian smelting site, containing scores of furnaces and storage pits for copper ore and fuel.*

In 1981 John Merkel, a graduate student at the London Institute for Archaeo-Metallurgical Studies, carried out an experimental copper smelt at Timna, in the Sinai desert in Israel, following precisely the methods used by the Egyptians in the second millennium BC. Merkel built his furnace near an Egyptian furnace which had been excavated in the 1970s, and copied its design. He used local clay for the lining and packed it around with stones and sand to help keep in the heat. The copper ore was malachite from the area, mixed with a flux of local iron ore, and the fuel was charcoal made from desert trees.

Merkel sets his furnace into a low slope.

Clay nozzles or tuyeres will deliver air.

Merkel's furnace has three tuyeres.

Each tuyere is supplied with air by a pair of 'pot bellows', based on those known to have been used by the Egyptians. A simple flap valve in the leather cover opens as the cover is pulled up and closes as the worker tramps down, thus forcing air through the pipe into the furnace.

The furnace is fed with a mixture of copper ore, charcoal, and iron ore to act as a flux. It is blown with the pot bellows for several hours, to maintain a temperature above the melting point of copper– 1084C. Copper is reduced from the ore and collects in the furnace bottom, beneath the lighter molten slag formed by the remaining materials.

A dipstick shows the depth of slag and copper.

After cooling overnight, the furnace is opened.

Threads of copper metal can be seen in the slag from the furnace. In a typical Egyptian smelt, the copper formed an ingot of pure metal in the bottom, while the molten slag was tapped off through a hole in the side of the furnace. Merkel's experiment, while not as successful as the routine accomplishments of the ancient Egyptian smeltermen, at least demonstrated that the process, as recorded in tomb paintings, does work.

ore flux, dressing areas where the ore was crushed, and scores of furnaces where the copper was reduced. Large heaps of slag beside the furnaces indicate the scale of production. The furnaces themselves, made from clay and packed around with stones and sand to retain the heat, were not much bigger than the nearby Chalcolithic furnace of two thousand years earlier, but there had been significant advances in technology.

The Egyptian furnaces were blown by two or perhaps even three tuyeres, or clay nozzles, each connected by a pipe to its pair of pot bellows. These were pottery bowls with a skin stretched across the top, and a cord attached to the centre of the cover. A man stood with a foot in each bowl and pumped air by tramping with each foot alternately, holding a cord in either hand to pull up the cover after each tramp. As he lifted the cover, air was sucked into the bowl through a flap valve in the skin. As he tramped with his foot the valve closed and the air was forced through the pipe into the furnace.

The application of forced draught to furnaces was a significant advance in smelting technology, for the higher temperatures produced a much better separation of molten copper from the slag. The slag, being lighter, floated on top of the copper. This enabled the smeltermen to get rid of the slag by tapping it off through a hole punched into the side of the furnace, just above the level of the copper in the bottom. The slag was allowed to run out into a shallow depression in front of the furnace. When the furnace had cooled the ingot of more or less pure copper was lifted out of the furnace bottom, and broken up and re-melted in a clay crucible to remove impurities. In such furnaces up to fifteen kilograms of copper could be produced with a single smelt. The series of photographs on pages 28 and 29 shows how John Merkel, an American graduate student at the London Institute for Archaeo-Metallurgical Studies, successfully smelted copper at Timna in 1981, using local ores and the same furnace technology as the ancient Egyptian smeltermen.

One of the many ways in which the Egyptians used their copper is recorded in a finely detailed series of tomb paintings which have survived for more than three thousand years beneath a rocky hillside near the Valley of the Kings at Luxor, on the Nile. The tomb is that of Rekhmire, vizier or lord chamberlain to King Thutmose III, who reigned during the eighteenth dynasty of the New Kingdom, in the second millennium BC. Rekhmire's areas of responsibility were broad, ranging from tax collecting to building construction, from law-giving to craft manufacture. All were meticulously recorded on the walls of his tomb, in scenes which are as fresh and clear today as when they were painted.

One scene shows in detail the casting of a pair of large temple doors. The raw metal arrives on the shoulders of workmen in two forms: a large slab of copper, and smaller ingots, presumably of tin, in baskets. Another workman delivers charcoal for the furnaces, each of which is blown by a pair of pot bellows, like those described earlier. The metal is shown being melted in clay crucibles over the furnaces. When the charge is molten two workmen lift the crucible with green poles,

recently cut and full of sap. This prevents them from catching fire just long enough for the metal to be poured into the mould, which stands on edge and has a number of funnels along the top. These act both as casting channels and vents for the escape of gases during the many pours that would be necessary to fill a mould of that size.

The Rekhmire tomb painting is one of the most explicit illustrations of early metal working that we have. It confirms that by the second millennium BC the technology of handling bronze on a large scale was well developed. However, neither this painting nor any archaeological evidence provides a clue to the greatest mystery of the Bronze Age: where did the tin come from, and how were its ideal properties for alloying with copper discovered?

Early smelted copper contained traces of arsenic, antimony, silver and lead because of their common occurrence in copper ores. The effects of their presence in copper would have been noticed, and perhaps even experimented with. But where tin is concerned there is one crucial difference: tin is seldom an impurity of copper ores. Furthermore, tin ores are rarely found in the same areas as copper ores. In fact, tin deposits themselves are comparatively scarce. (This makes tin, even today, an extremely expensive metal; in the five years to 1983 its price varied between $12,000 and $16,000 per tonne, eight to ten times the price of copper or aluminium.)

Tin generally occurs as tin oxide, or cassiterite, in the form of veins in hard rock, usually granite. Where it becomes exposed it may break down and wash away into stream beds. There it resembles a heavy, black gravel, and must be panned or dredged like gold. This is the only way that early man could have obtained tin, because he lacked the means of extracting it from granite rocks.

Above left: This Egyptian tomb painting, dating from the second millennium BC and showing the melting of copper for the casting of a pair of large temple doors, is one of the earliest records of metal working. The furnaces are being blown by pot bellows. *Above right*: Another section of the painting shows workers pouring molten metal into the mould, while others arrive with ingots to be melted. *This painting is in the tomb of Rekhmire, Lord Chamberlain to King Thutmose III, one of the Pharaohs of the 18th Dynasty of the New Kingdom. The tomb is in a hillside near the Valley of the Kings at Luxor.*

31

The world's major tin deposits are in Malaya, China, Thailand, Bolivia, Nigeria, Saxony-Bohemia, Australia, Zaire and Cornwall. In ancient times there were smaller deposits – now essentially mined out – in Sardinia, Italy, Spain and possibly Afghanistan. In none of these areas, however, is there any evidence of the early phases of bronze making. And there were no known tin deposits in or near any of the great bronze-making centres of the Near East and the eastern Mediterranean. How, then, did the Bronze Age get under way?

The island of Cyprus expresses the dimensions of this question, although it leaves all the answers up in the air. By about 2000 BC, like the other major centres of culture in the Near East, Cyprus had entered its full Bronze Age. Tin bronze had displaced arsenical copper as the

metal of choice for tools, weapons, utensils, and works of art such as sculpture. The island had become a key centre in the bronze trade, not just because it lay across the main sea routes of the eastern Mediterranean, but primarily because of its immense deposits of copper. The great copper mines at Skouriotissa, in the hilly centre of the island, have only recently closed down after thousands of years of operation. The modern name itself comes from the Roman term for copper, *aes cyprium*.

Cyprus was quite possibly the source of one of the most distinctive artefacts of the Bronze Age, the copper 'ox-hide' ingot. This was a slab of almost pure copper, weighing up to thirty kilograms, and cast in a characteristic shape like a dried, stretched ox hide, with two projecting 'arms' at either end. The shape may have been designed to make the ingot easier to carry by two people, or by one man on his shoulder. Such ingots have been found in various parts of the Mediterranean, and in ancient shipwrecks. Since Cyprus was noted as an exporter of copper (it is thought to have been the legendary 'copper island' of Alashia, mentioned in ancient texts), and since the bull was much revered in the island as a symbol of power, the people of Cyprus might well have considered it appropriate to cast their main source of income in a form derived from their sacred animal.

Excavations at many sites on Cyprus have established the scale of the copper and bronze industry on the island. One of the most recent discoveries was the ancient city of Kition, found beneath the suburbs of the modern coastal township of Larnaca. Beginning in 1962, archaeologists uncovered a great array of temples, palaces, courtyards and columns. Next to the temples, and connected to them by doorways, was a complex of copper-smelting workshops, littered with slag and other evidence of metal working. The proximity of the workshops to the temples, and the many finds of exquisite copper and bronze artefacts among the buildings, underline the importance of the industry and its high place in Cypriot Bronze Age society.

In one sense, however, the significance of the Cyprus bronze industry may be the fact that it existed at all, despite the total absence of any sign of tin on the island. Like the great centres of Near East bronze making at the time, it appears to have been totally dependent upon some external supply of tin. The source of this tin in the ancient world has been the subject of endless speculation and dispute among archaeologists, prehistorians, and now archaeo-metallurgists. One who has devoted a great deal of study to the problem is James D. Muhly, Professor of Ancient History at the University of Pennsylvania in Philadelphia.

'Archaeologists, like nature, abhor a vacuum,' states Muhly. 'Tin is a major vacuum in the understanding of the transition from the age of stone to the age of high energy plastic materials such as metals . . . because the geology of its exploitation remains to be explained. The data are consistent and they present us with a geological vacuum at the present time. [I] have combed the available data on tin, visiting alleged sites as distant as Kuhbanan in southern Persia and Mokur in Afghanistan. There are enough analytical evidences available at present to warrant the belief that in the early Bronze Age the Black Sea mountains and the Zagros mountains north of Hamadan afforded alluvial tin as they afforded alluvial gold – and that tin was quickly exhausted. In relatively quick order tin was found in more distant sites, stretching from Hungary to Spain and possibly to Thailand. But there is no evidence at present to suggest that such exotic sources played any role in the trade and metallurgy of the Near East Bronze Age.'[10]

Muhly points out that the absence today of alluvial tin from river beds in the Near East does not necessarily mean that no tin ever existed there. Alluvial tin is like alluvial gold, in that it can be completely panned out of a stream. However, such a process does not affect the 'mother lode' of tin that must exist somewhere nearby, in the form of veins in the rocks. Unfortunately, no such tin-bearing formations have been found near the major centres of Bronze Age civilisation, despite exhaustive searches in once favoured areas such as the mountains of Iran, Anatolia and the Caucasus. Soviet geologists reported in 1981 the discovery of tin-bearing rocks in occupied Afghanistan, but no details were provided.

One expedition, led by Muhly, did make a surprising find of tin in a totally unexpected area: Egypt. Significant tin deposits are now thought to exist in the desert north-west of Aswan, based on the discovery of alluvial tin in a number of dry river beds. There is no evidence, however, that these deposits were worked or even known in the Bronze Age. The Egyptians themselves did not begin making tin bronze until later.

Meanwhile, an intriguing discovery in 1976 might hold the clue to the origin of tin supplies for the Mediterranean bronze-making centres, although so far it has defied explanation. This was the appearance of what may prove to be the very first ingots of tin known from the Bronze Age.[11]

The story began with 'Arabian Nights' overtones, when ancient-looking ingots of metal began turning up in scrap metal shops in Haifa, the port in Israel. Several ingots of both copper and tin were eventually recovered from the markets by the Haifa Museum authorities. Israeli archaeologists traced them to an Arab fisherman, who said he had salvaged them from a wreck off the Israeli coast. Over about a year he had raised and sold several tonnes of metal, but had then lost the wreck under shifting sandbanks.

The ingots were intensively but inconclusively studied for evidence of their origins by the British Museum Laboratory, the Max Planck Institute of Atomic Research in Heidelberg, and the Museum of the University of Pennsylvania in Philadelphia. What makes the two tin ingots particularly interesting is that they carry engraved marks which closely resemble the Cypro-Minoan script. This was used on Cyprus and in Syria during the Bronze Age, between 1500 and 1100 BC, but also in Spain much later, around 700 BC.

Since it is known that Cyprus had no tin, the ingots could hardly have originated there. But the island might have been a distribution centre for tin brought from somewhere else, and supplied, with Cypriot copper, to users of bronze such as Egypt. Just such a mixed shipment might have been on its way along the Palestine coast towards Egypt when the ship sank. Another explanation is that the tin and copper was being shipped home from Spain, at a much later date, by the traders from the Levant coast who later became known as Phoenicians. Whatever the source or destination of these mysterious ingots, their identification would help to fill the 'vacuum' that Muhly has referred to.

The conclusion which most authorities are now coming to accept is that the true Bronze Age in the Near East and eastern Mediterranean began when, and only when, a far-flung and fairly reliable network of trading contacts had been established across the ancient world. Extending in the west as far as Spain and perhaps Cornwall, and in the east at least as far as India (and possibly further, as will be suggested), these routes were the major arteries of supply for the tin which was in such demand in those early centres of civilisation.

Long journeys in those times must have been slow and dangerous, and the tin supplies may have been interrupted for quite long periods. There is evidence for this in that during the Late Bronze Age, there were occasional reversions to arsenical copper in both Mesopotamia and Egypt, as if the materials for making true bronze were not available. (The final collapse of the Bronze Age, described later, is thought to have been linked to such interruptions on a massive scale.)

As far as the bronze-using cultures of the eastern Mediterranean were concerned – Cyprus, Crete, Mycenae and Greece, and perhaps Egypt – all references in the classical texts are to tin supplies from the west, from sources which included the 'Cassiterides'. These were an unidentified group of islands reputed to lie in the Atlantic, but which some historians think may have been Cornwall, where it is known the Phoenicians often called. There was certainly plenty of tin in Cornwall, Spain and Bohemia, but there is no firm evidence that it was ever recovered during the Bronze Age. It is possible, although unlikely, that the limited tin resources of Italy and Sardinia were sufficient for the Mediterranean bronze industry.

When it comes to the great bronze-producing centres of Mesopotamia, Assyria and Anatolia, the presumption of an extremely distant source of tin is, on present knowledge, unavoidable. All textural and archaeological evidence points to a source in the east, and a trade in tin on a very large scale, but its origin remains a complete mystery.

As Muhly puts it: 'We do not have a single clue regarding the source of the tin used in Anatolia during the third millennium BC. Nor do we really know the source of the tin that the Old Assyrian merchants brought into Anatolia. Yet the extent of that trade is truly staggering. A group of one hundred and eighty-nine Old Assyrian letters, out of the many thousands that are known, deals specifically with the caravan trade. They show that, within a period of about one hundred years at the beginning of the second millennium BC, ninety donkey caravans brought almost eleven tons of tin from northern Mesopotamia into central Anatolia. This is trade on a major scale, not trade resulting from the casual chance find of a small amount of tin. There must have been some well organised source or supplier that could be counted on to produce consistent amounts of tin year after year.'[12]

To the east, the nearest known sources of tin are enormous distances away: Malaya, Indonesia, Thailand and China. However unlikely it may seem, therefore, we must consider the possibility that tin from Southeast Asia was carried thousands of kilometres to the Near East as long ago as the fourth millennium BC. From all accounts the overland routes to the Far East were not opened up until thousands of years after this, but it is possible that small coastal vessels made their way along the coasts of Malaya, Burma and India into the Persian Gulf, where their cargoes could have been unloaded and carried overland.

There is one quite startling corollary to such a source of tin for the Near East Bronze Age. Recent discoveries in South-east Asia have opened up the possibility that the secret of bronze making itself may have been discovered in the East and carried with the tin to the Near East. Such an iconoclastic idea goes against all long-held theories of cultural and technological 'diffusion' from the 'cradle of civilisation' in the Near East. But belated investigation of vast regions hitherto virtually ignored by Western archaeology and prehistory is forcing it into consideration.

The new chapter in the history of metallurgy is being written in South-east Asia, in the region once known as Indo-China. This vast expanse of jungle-covered mountains and valleys, cut by great rivers and dotted with open, grassy plateaus, remains one of the least-known areas of the world – at least as far as archaeology is concerned. Traditionally regarded as a kind of cultural backwater between China and India, it has attracted little attention, even from those interested in the origins of civilisation in Asia.

The present populations of this region live an agrarian existence, and there was no evidence that their predecessors had ever experienced an original Bronze Age, like that which marked the rise of civilisation in the West. There had been scattered finds of bronze as well as iron objects, but these had not been properly dated, and were considered to belong to an Iron Age period, around the beginning of the Christian era. The apparent coexistence of bronze and iron suggested an importation of technology at a comparatively late date. The earliest metal artefacts of any real archaeological significance was a collection of bronze drums, found at Dong Son in what is now northern Vietnam. These were dated to the last few centuries BC, and were presumed to have come from the north, perhaps from China.

Only one aspect of South-east Asian prehistory had aroused much curiosity in the West – the suggestion, floated at intervals since the nineteenth century, that this area might have been the origin of a number of important food plants that are now cultivated widely throughout the Indo-Pacific region. These include rice, yams, taro, sugar-cane, coconuts and bananas. Some botanists were even beginning to wonder whether we should be looking for the earliest manifestations of food-growing in South-east Asia, with its tropical climate and its profusion of plant forms, rather than in the climatically more austere environment of the Near East.

So it happened that in 1965 a somewhat unorthodox archaeologist at the University of Hawaii, Professor Wilhelm Solheim, took two of his students, Chester Gorman and Donn Bayard, to look in South-east Asia for evidence of early societies which might have experimented with agriculture. At the time, a dam on the Mekong River was being planned in Thailand, near the border of Cambodia, and the opportunity arose for Solheim and his students to survey the area before it was flooded. They found a likely-looking mound, and Donn Bayard excavated it in 1965 and again in 1968. Gorman, meanwhile, went off on his own and excavated a cave on higher ground, to obtain material for his doctorate.

36

In what he named Spirit Cave Gorman found not only flint tools and other signs of Stone Age occupation dating back to about 10,000 BC, but seeds and food remains hinting at the cultivation or at least the propagation of food plants. This find alone would have been enough to bring South-east Asian prehistory to world attention. But at the first excavation site, called Non Nok Tha, Bayard and his team had made a much more startling discovery: bronze axes and socketed tools, bronze bracelets, and sandstone moulds and clay crucibles used for bronze casting. The dates, when established by carbon dating and published, reverberated like an archaeological landmine.[13]

Here in northern Thailand, it appeared, there had existed in the fourth or third millennium BC a hitherto unsuspected society which had developed a bronze technology as early as any of the accepted centres of innovation in the Near East, Europe, India or China. There was incredulity, even disbelief, at dates of 3000 to 2300 BC for tin bronze in this region. The site itself had proved difficult for Solheim and his team to evaluate with precision, consisting as it did of some four thousand years of successive burials and their accompanying grave

Thailand, which had always been considered a technological backwater, lacking any significant Bronze Age tradition, was the scene in the 1960s of a series of archaeological discoveries which challenged the established theories of cultural 'diffusion'. *Temple at Ayutthaya, ancient capital of Thailand, on the plain north of Bangkok.*

goods, intermingled in less than two metres of deposit. What was needed was more evidence, from deeper and more accurately datable deposits. This came, overwhelmingly, in the early 1970s, when attention was focused on a small village on the Khorat Plateau in northern Thailand, called Ban Chiang.

For some years previously the villagers of Ban Chiang had been unearthing painted pots of unusual design while gardening or digging post-holes for their houses. The finds attracted no attention until 1966 when the village was visited by Stephen Young, a student at Harvard and the son of the United States ambassador to Thailand. Young found pots being exposed by erosion along a roadway, and realised that Ban Chiang was built on some kind of archaeological site. He took a few pots back to Bangkok and showed them around.

Through the aroused curiosity of Elizabeth Lyons, a fine arts consultant at the American embassy, some of the pottery found its way back to the United States. At the University of Pennsylvania Museum in Philadelphia it was examined and dated by the technique of thermo-luminescence. This is a useful guide to the age of fired pottery, although not very precise. Meanwhile, the Thai Department of Fine Arts carried out a small excavation at Ban Chiang, and found not only more pots but items of corroded metal.

When the pottery dates were revealed in Philadelphia they were even more unbelievable than those from Non Nok Tha; they were around 4500 BC. And objects apparently made of bronze had been found at the

Excavations in the village of Ban Chiang in northern Thailand unearthed rich burial sites of a hitherto unsuspected culture, which flourished for nearly four thousand years, from around 3600 BC to about AD 300. *The graves contained not only painted pots, but axes, bracelets, neck bands, and other decorative objects in bronze and iron.*

same site! It was enough to set in motion a major project by the University of Pennsylvania Museum, in collaboration with the Thai Fine Arts Department. The team was headed by Chester Gorman and Pisit Charoenwongsa. They arrived with their helpers in Ban Chiang in 1974.

The village turned out to have been built on a large, low mound, only slightly raised above the level of the surrounding paddy fields. By the time the excavation team was ready to start work, however, most of the open spaces in the village had been dug up by villagers looking for items for the flourishing black market in pots and bronze artefacts which had been created by dealers and tourists. One of the few undisturbed areas was a village street, and here the archaeologists began their excavation. They had immediate and remarkable success.

In a space no greater than two hundred square metres in area and five metres in depth they found one hundred and twenty-three burials, covering a time span of nearly four thousand years, from 3600 BC to about AD 300. The graves contained not only the skeletons of adults and children, but a profusion of grave goods. These included hundreds of painted pots and many items of bronze and iron, ranging from axes and spear heads to bracelets and other decorative trinkets.

The eighteen tonnes of material were distributed to laboratories and universities in various parts of the world, to establish when the artefacts were made, and to find out more about the early inhabitants of Ban Chiang. The dating was complicated by the fact that, as at Non Nok Tha, later graves were sometimes cut down into lower and therefore

The most startling finds at Ban Chiang in Thailand were the bronze objects, because no bronze tradition of such antiquity had been known previously from anywhere outside the 'cradle of civilisation' in the Near East. *Left: The earliest find was this bronze spear head, dated to around 2000 BC. Above: A set of finely cast bronze bracelets. These and other finds, on loan from the Thai Fine Arts Department, are being studied at the Museum of the University of Pennsylvania, Philadelphia.*

One of the burials excavated at Ban Chiang was this large adult male, with bronze bracelets on his left wrist, and a socketed bronze axe head near his left shoulder. *Burial 25, given the name 'Vulcan', dates from 1600 to 1200 BC.*

older burials. This meant that some objects were found in levels that suggested a much earlier date than was eventually established for them. However, the dates for the metal artefacts finally released by the University of Pennsylvania Museum in 1982, although later than Gorman's preliminary dates, are profound in their implications. (Chester Gorman himself sadly died before he could revise the dates himself.)[14]

The excavations at Ban Chiang show that by about 2000 BC some people in this remote part of northern Thailand were making sophisticated objects in bronze, with near the optimum tin content of ten per cent. They showed an excellent knowledge of alloying, casting, annealing and working of bronze. Furthermore, by about 1200 BC these hitherto unknown metal workers were making use of two metals in one object. Among the finds was a spearhead with a blade of forged, smelted iron, on which a bronze socket had been cast to take the shaft. Bi-metallic objects of this age are extremely rare, from anywhere in the world.

About 1200 BC, in the stratigraphy of Ban Chiang, iron tools and utensils appear beside the bronze. With the passing centuries iron becomes more common, while bronze reverts to purely decorative uses. The last dates, from the upper levels of the excavation, are from around AD 300. After that the Ban Chiang cultural tradition apparently disappeared. The site of the once busy settlement remained uninhabited until about two hundred years ago, when the ancestors of the present villagers migrated into the area from Laos, to the north, and cleared the land and began the present phase of farming.

The questions raised by the discoveries at Ban Chiang are many, and some remain unanswered and perhaps unanswerable. As one of the Americans who excavated the site puts it: 'What we found here is like finding hub-caps under the ruins of ancient Rome!' One major question is this: what were the factors which enabled such a technically advanced society to emerge from the hunter-gatherer Stone Age culture, called the Hoabinhian, which at that time existed throughout South-east Asia?

A faint picture of that distant period is beginning to form as a result of the growing body of archaeological, anthropological, botanical and linguistic work in this region, especially by the Americans. It now seems that about 4000 BC a wave of settlers from the swampy coastal plains of South-east Asia moved up on to the comparatively dry Khorat Plateau of Thailand, and began following a new kind of existence. They still lived mainly on wild foods and animals, but they were beginning to establish a more settled, food-producing way of life. They built light bamboo houses, made pottery, and kept cattle, pigs, and chickens. More significantly, they had begun the process of domesticating and cultivating rice.

Ban Chiang itself was apparently first settled about 3600 BC, although the first metal objects do not appear until about 2000 BC. What is puzzling about this, however, is that the bronze tradition, when it does appear, seems to have been born almost fully developed, without any preceding copper culture such as we find in the Near East and elsewhere. A few of the objects from Ban Chiang and Non Nok Tha are low in tin, and cannot be described with certainty as 'deliberate' bronze. But in general there is a striking absence of copper or early experimental bronze.

Could the tradition of bronze casting, or even the objects themselves, have been introduced into the area from somewhere else? No cultural affinity with any external source has been found. It would be unusual for a people to borrow or import a tradition such as bronze making without at the same time adopting other recognisable cultural traditions

Support for the belief that the metal objects found at Ban Chiang were locally made and not imported is provided by this bi-valve stone mould, used for casting bronze axe heads. *The axe head shown was cast from the mould at the University of Pennsylvania Museum.*

in, for example, pottery. As for the metal artefacts, the finding of many clay crucibles, some still containing droplets of bronze, strongly suggest local production. No actual smelting sites were found, but these may well have been located near the deposits of both copper and tin in the mountains around the rim of the Khorat Plateau.

Against this background, the earlier discoveries of undated bronze objects in South-east Asia take on a new significance. In a paper presented at the UNESCO symposium on bronze culture in Bangkok in 1976, Pisit Charoenwongsa pointed out that tin is found in rich alluvial deposits all the way from southern China down through Thailand and the Malay peninsula into Indonesia, while copper has been found in all those countries except Malaya. On the face of it, there appears to be a greater probability of the discovery of tin bronze being made in this region than in the Near East, where copper and tin do not occur together.[15]

However, the Ban Chiang bronze culture appears to depart from the pattern of development which accompanied the Bronze Age elsewhere, particularly in the Mesopotamian city states, and which seemed to provide the proper social context for the regular, organised production of metal artefacts. The difference is defined by Joyce White, a former student of Chester Gorman's in Thailand, and later curator of the exhibition of Ban Chiang material mounted in 1982 by the Pennsylvania University Museum, the Smithsonian Institution and the Thai Fine Arts Department. In the catalogue White writes:

> That the technological precocity found in Ban Chiang occurred in simple village contexts that derived their subsistence from hunting, gathering and simple agriculture is most intriguing. No urban, state or military stimulus from within or without the region is in evidence. No complex, stratified social organisation appears to have been a cause or a consequence of the development of metal technology. In fact, what we seem to be finding is the outline of a peaceful Bronze Age unlike that experienced anywhere else in the world. The bronze and iron artefacts reflect non-military, non-urban applications. There are no mace heads, swords, battle-axes or daggers. The spear and arrow heads are few, and might well have been intended for hunting. The majority of the items found are, even during the later periods of high output, for decoration and adornment. Nor does the possession of metal goods appear to symbolise the possession of wealth or status; more children than adults were found with bronze and iron bracelets.[16]

Ban Chiang may therefore be regarded as a child of its revolutionary times, encouraging us to look for the origins of metallurgy not only in urban societies with their economic resources but also, perhaps, in much simpler communities, where need or the absence of alternatives may have stimulated human curiosity and imagination. As Joyce White concludes in her essay on the Ban Chiang exhibition: 'Metals do not create cultural development, people do. Whatever the source or influence of the ancient metallurgy of South-east Asia, it can no longer be denied that special and economic development in the region from prehistory onwards had a profound impact on far-flung regions of the world.'

42

Certainly there is today a growing body of evidence which firmly identifies South-east Asia as the nucleus of the great expansion of people, culture and language across the entire Indo-Pacific region, from Madagascar in the west to Easter Island in the east, which began around 4000 BC. All the varied peoples which inhabit this region have in common more than the elements of some ancestral Austronesian language; they share many domesticated food plants, including coconuts, bananas, breadfruit, taro, yams, arrowroot and sugar-cane. It now emerges that South-east Asia still contains relatives of all the differentiated Austronesian languages of the Indo-Pacific region, as well as the nearest wild relatives of those widespread food plants.

In seeking causes for that expansion of people – one of the greatest migrations in human history – a few archaeologists are beginning to speculate that economic or commercial interests of some kind might have been responsible. The impetus could have come from a developing region of wealth and population on the Asian mainland which provided growing markets for products from abroad. Ban Chiang may well be the first archaeological evidence of such a culture. It would not have been the centre itself, but perhaps an outpost of a power whose even more substantial remains may lie unsuspected beneath the jungles and paddy fields of South-east Asia, as Angkor did until 1861, and Ban Chiang itself did until 1974.[17]

However, besides the many fascinating developments in this story are others which are equally intriguing, because they did *not* happen. One that stands out concerns Ban Chiang itself. That culture, despite its initial technological impetus, eventually failed and disappeared. In the words of Chester Gorman and Pisit Charoenwongsa: 'Why did an expandable, grain-based agricultural system in the context of an innovative, metal-using culture not set a trajectory leading to higher levels of political organisation or urbanisation?' That is an intriguing question without an answer – except for the possibility that Ban Chiang may have been, like Catal Huyuk, an example of the 'false start'.[18]

By contrast, there lies to the north of Ban Chiang a vast country where innovation and invention did take root, to a degree which is only just beginning to emerge from thousands of years of self-imposed isolation. China is finally yielding evidence which is overwhelming the formerly unassailable theory of diffusion, at least where metallurgy is concerned.

In Henan province, in central China, the scene today can have changed little in the past four thousand years. The fields are dotted with mud-brick villages, and stretch in an orderly brown and green patchwork to the misty horizon. People move along the furrows, hoeing and weeding their crops. Animals pull ploughs and carts piled high with vegetables. The fertile topsoil is in places hundreds of metres deep; it is the wind-blown loess from the Asian highlands. This is the flood plain of the Huang-Ho, or Yellow River. It is one of the first places where Neolithic man began to settle down and cultivate his food, and one of the great cradles of civilisation.

The Huang-Ho has been called the River of Sorrow, because of its history of floods. These have been responsible for some of the most devastating calamities in China's history, and they have also buried some of its most gifted triumphs. Unforgettable evidence of this has been provided by the excavation of the ancient city of Yin, which lies beneath wheat and cotton fields just outside the modern industrial city of Anyang, some twenty-five kilometres from the present course of the Huang-Ho.

Yin was first settled around 1600 BC by a people called Shang, moving from an earlier site at Zheng-zhou, to the west. In 1400 BC the Shang emperor, Pa'anken, established his capital at Yin, and for the next three hundred years the Shang dynasty flourished. Its people must have lived much as the villagers around Anyang do today, but some among them also found time to create, in bronze, one of the most powerful and accomplished bodies of art in history.

An intriguing aspect of the Shang story is that the bronzes were known and admired all over the world long before there was any archaeological evidence of the Shang dynasty itself, or of its capital, Yin. This unusual situation had come about largely because the isolation of China from the outside world had prevented the development of the art and science of archaeology. And there was another reason – the very nature of the objects themselves. They were not tools or weapons or utensils in daily use, which would have been relatively common, and therefore plentiful in ancient sites. They were ceremonial vessels, produced in restricted numbers, for the offering of food and wine to ancestral spirits. These offerings were the core of the sacrificial rituals performed at intervals by the emperor and the nobles, and the bronze vessels were eventually buried with them in their graves.

Over the three thousand years after the collapse of the Shang dynasty in 1122 BC, many of the graves around Yin were looted and their contents dispersed. Quantities of the bronzes were undoubtedly sold and melted down for scrap metal. But some found their way into the hands of visitors to China, such as missionaries and soldiers of fortune, and eventually into private collections and museums abroad. By the beginning of this century the Shang bronzes had established their place among the great artistic and technical achievements of early civilisation.[19]

It was not until the late 1920s, however, that the first scientific excavations were begun on the site of the Shang capital, Yin. Work continued, interrupted for long periods by the war against Japan and then the Revolution, until the 1950s. Then, after the full extent of the ancient city had been revealed and recorded, the archaeologists reburied the ruins and returned the fields to cultivation. But from undisturbed tombs and graves they had recovered even more striking examples of Shang bronze than those upon which its fame had grown. The result is that China itself, in its museums, can now match the best foreign collections from this most creative period in its history.

Shang pieces were made from high quality bronze, containing from five to thirty per cent tin and two to three per cent lead (which made the bronze flow more easily). During its long interment the bronze has acquired a distinctive green patina, through surface oxidation of the copper. This finish is greatly prized by collectors, and forgers of Shang

44

bronzes have tried every conceivable method of reproducing it. There is an account of one family in China which went about the business with an admirable sense of continuity; each generation buried Shang copies in damp, acidic soil, to be dug up and sold by the next generation but one.

There are at least thirty main styles of Shang ritual vessels, ranging from wine cups in various shapes, called *chueh* or *chia*, to huge cauldrons on stout legs, called *ting*, weighing the best part of a tonne. In between are all kinds of bowls, steamers and dishes for preparing and serving offerings during the rituals.

Apart from their unusual shapes, what makes the Shang bronzes immediately recognisable is their decoration. They carry relief patterns of extraordinary richness, many based on stylised representations of

The products of the Shang bronze tradition, which flourished in China between about 1400 and 1100 BC, range from small vessels of grace and delicacy to this massive cauldron, used in rituals, and buried with one of the Shang rulers. It weighs 875 kilograms, and is the largest casting in metal anywhere in the world which can be dated to the second millennium BC. *Found at Anyang in 1939, it is now in the National Historical Museum, Peking.*

45

tigers, water buffaloes, elephants, owls, fish and mythical dragons. The decoration is often integrated into the design of the vessel in masterly fashion, with the legs of an animal forming the feet, and the head the lid.

The outstanding technical achievement of the Shang craftsmen was that all their products, regardless of their size, shape or function, were made by casting. They made no use of Western metal-working techniques of hammering, chasing, raising or incising. They were masters of molten metal, and they cast each piece complete in every detail, although sometimes in sections which were later assembled.

The starting point was a clay model, on which all the decoration had been stamped or carved. This was baked hard, and from it a clay mould was taken and baked in its turn. When filled with molten bronze it yielded the finished object with all its rich decoration perfectly reproduced. Legs and handles for the larger cauldrons were cast separately and attached with molten bronze.

Modern metallurgists are astonished by the technology of the Shang bronze casters, and particularly by such examples of their skill as the huge *ting* in the National Historical Museum in Peking. This massive rectangular cauldron on four square legs was unearthed by villagers near Anyang in 1939, and used for some years by them as a store for pig food. Although not as aesthetically pleasing as some of the smaller Shang pieces it has one unchallengeable claim to recognition: it is the largest surviving casting in metal anywhere in the world that can be firmly dated to the second millennium BC. It weighs 875 kilograms, and its body was cast in a single piece – a feat which would tax many metalworks today.

Recent excavations have provided some clues as to how the Chinese, around three thousand five hundred years ago, could have cast nearly a tonne of molten metal. At Anyang, archaeologists have found fragments of crucibles big enough to melt such quantities of bronze. In some cases multiple and simultaneous pours were obviously used. Clay channels, found on the floor of a foundry, lead from separate furnaces and crucibles to a central point where the moulds were filled. The channels were set in beds of burning charcoal, and the moulds themselves were pre-heated to ensure that the bronze flowed freely. Such evidence argues that in China, in the second millennium BC, there existed a metal technology which approached the scale of an industry. The question is, where did it have its origins?

Until recently there was little reason to dispute the accepted view that Chinese metallurgy was, in a world historical context, a comparatively late starter. The basis for this belief was the 'sudden' appearance of the Shang bronze tradition, fully developed and technically advanced, around 1400 BC. Since there was no evidence of the familiar progression from native copper to smelting, and from arsenical bronze to tin bronze, and since the technology of tin bronze was already well established in the Near East by the second millennium BC, the obvious explanation of the Shang 'phenomenon' was that the knowledge of bronze making had reached China from the West.

Today that theory has been shaken, if not shattered, by a series of discoveries in the one area that had been ignored by archaeologists in the past, and denied to them in more recent times: the vast expanse of China itself. A most comprehensive interpretation of this new work has been made by Noel Barnard, Senior Fellow in the Department of Far Eastern History, Institute of Advanced Studies, Australian National University. After a lifetime spent in Chinese studies from original sources, Barnard is one of the leading Western authorities on Chinese metallurgy. And he is in the forefront of those who argue that the evidence is now very strong for the independent discovery of metallurgy in China, including the smelting of ores and the alloying of copper and tin to make bronze.[20]

Barnard points out that one major revision of views is called for in relation to the apparent lack of a pre-bronze metal-working tradition. In fact, many sites in central China, and in particular a place called Erhlitou, not far from Anyang, have now yielded bronze artefacts showing a much more rudimentary development of metallurgy. Some have such a low percentage of tin as to be classified as copper, or at least 'accidental bronze'. Barnard also quotes a discovery by a Japanese archaeologist, Wajima Seiichi, made in China in 1942 during the Japanese occupation but not published until 1962. Seiichi found, in a Lungshan 'black pottery' site, dated to around 2000 BC, a piece of coarse earthenware with a layer of copper slag adhering to it. Analysis confirmed that the slag was the result of copper smelting.

Barnard publishes maps and tables of all sites in China where metal artefacts had been found up to 1975, and matching maps showing all known deposits of copper, tin and lead. The archaeological evidence shows a pronounced centrifugal pattern of metallurgical development, centred on the Anyang area, overlapping and in fact matching the pattern of mineral distribution. The greatest concentration of ore deposits lies within a radius of three hundred kilometres of Anyang. The earliest bronze-using sites are also clustered in this same core area, in the Huang-Ho basin. Move outwards in any direction – even to the west and south, the once presumed routes of entry of metallurgy – and the dates for bronze become later.

To Barnard this suggests not the diffusion of knowledge from the West into China, but from the centre of China to its more distant regions. As for the origins of Chinese metallurgy, Barnard considers that they can be found in the highly developed pottery industry of the immediate pre-metal period. Much new evidence for this proposition has been published in the past twenty years, as a result of the steadily expanding programme of archaeological excavation in China.

It has always been known, for example, that Chinese potters had more advanced kiln designs and a greater command of temperatures than their contemporaries in the West. Hence the invention in China of porcelain, which is made from a type of clay found all over the world, but which requires a firing temperature in excess of 1300°C. Tests show that pottery kilns found intact in Neolithic sites such as Ban Po, near Xi'an, dated to the sixth millennium BC, were capable of maintaining temperatures as high as 1400°C. This is more than sufficient to smelt copper, even without forced draught.

The Chinese bronze tradition declined steadily after the collapse of the Shang dynasty, but during its last phase, in the Eastern Han period (AD 25 to 220), it produced this undoubted masterpiece: a flying horse, balanced on the back of a swallow. 'Horse and Swallow', height 34 cm, found in the tomb of General Zhang in Gansu Province, now in the Gansu Provincial Museum, Lanzhou.

More significantly, perhaps, the casting and decorating techniques of the Shang bronze makers are clearly derived from the earlier pottery-making tradition, with its clay models, stamped and incised decoration, and earthenware piece moulds. There is also a direct continuity of form in many of the Shang bronze vessels from pottery vessels dated to 4000 BC and earlier.

One final argument for the independent origin of metallurgy in China is, paradoxically, a negative one. This is the absence from early Chinese artefacts of any sign of many metal-working techniques that were universal in the West: annealing, stamping, raising, sinking, engraving, inlaying and the use of sheet metal. Barnard argues that had there been any introduction of metallurgical knowledge into China before the second millennium BC, and the great flowering of the Shang period, it would almost certainly have involved these techniques of smithing. It may also be significant that there is no evidence that the native metals, gold, silver, or meteoric iron, were used in China until well after bronze was established, whereas in the West the opposite was true.

A late contribution to this discussion, not yet published in the scientific literature, was made in 1982 by Professor Tsun Ko during a visit to Australia to lecture at the Australian National University. Professor Ko is Vice-President (Research and Graduate School) of the University of Iron and Steel Technology in Peking, and a member of the Technical

Science Division of the Chinese Academy of Sciences. He told me of a number of recent finds and studies which suggest that metallurgy in China developed along quite different lines from the rest of the world.

One discovery was that two different kinds of bronze were used in China; one the standard alloy of copper and tin, and the other an alloy of copper and lead. The latter was developed in Honan province, where there was a shortage of tin. Professor Ko does not believe that this happened anywhere else in the world.

Another unusual find was of a piece of brass dated to about 2000 or 2200 BC. Brass is an alloy of copper and zinc, and it did not come into use in the West until long after other copper alloys. The Chinese brass was found in Shantung province, near the home of Confucius. Near by there were deposits of zinc-bearing copper. Professor Ko considers this find to be 'a natural smelting product from the local ore', possibly accidental.

Another difference from the West is the complete absence of arsenical bronze. Professor Ko said that Chinese metallurgists had recently analysed more than three hundred pieces of bronze, dating back before 1600 BC, and found no trace of arsenic in them.

Finally Professor Ko described the recent find of a knife in Gansu province, made of eight per cent tin bronze, which was firmly carbon dated by the millet seeds with it. The date was around 2800 BC. In the same area the archaeologists found copper-smelting products, including slag, malachite and corroded copper metal. The bronze is some seven to eight hundred years later than the first tin bronzes in the Near East, but Professor Ko considers it impossible for the knowledge to have travelled to China in that time.

Taking all these factors into account, it is now possible to argue, as Barnard does, that metallurgy began independently and followed its own course in China, and that a Bronze Age developed not much later than in the West. It would have grown out of a ceramic tradition, with casting virtually its sole technique. Because copper was difficult to cast, the Chinese metal workers appear to have very quickly devised an alloy, using readily available tin, which performed much better. Thus was ushered in one of the more obscure but most dazzling periods of the Age of Bronze.

But even as the Bronze Age reached its zenith across the ancient world, from the Atlantic to the Far East, its reign as the most useful and accommodating of metals was about to be abruptly ended. A series of powerful but still unclear events would force mankind to knock on another, more formidable door, and enter the Age of Iron.

CHAPTER 3
Bronze gives way to iron

The Qutub Minar, on the outskirts of Delhi, is a tall, tapering, fluted tower built of warm yellow stone. It rises to a height of seventy-three metres, and is considerably taller than the Tower of Pisa. It is in truth a tour de force of Mogul architecture, rivalling any freestanding pre-industrial structure anywhere in the world. For almost a thousand years its ornate balconies and inscriptions from the Koran have survived wars and conquests, lightning strikes and earthquakes, and it stands today as straight and true as when it was completed in the eleventh century AD. It has become a symbol of Delhi and the reign of the Rajput kings.

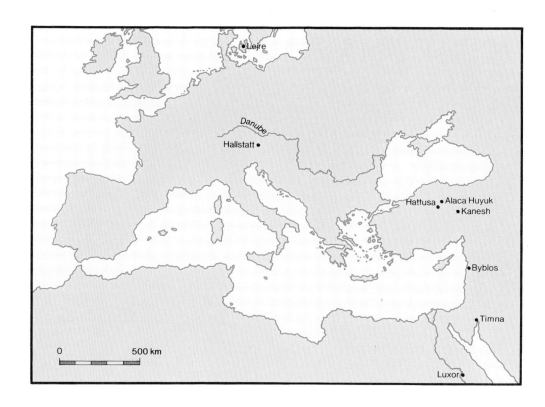

In a courtyard at the base of the Qutub Minar, however, there stands an even more remarkable example of ancient technology: the Iron Pillar. This smooth, massive column of black iron, seven metres tall and crowned with a stylised, finely wrought iron lotus flower, is polished to head height by the hands of pilgrims and admirers. It weighs six tonnes, and scientific analysis has shown it to be made of iron of extraordinary purity, averaging 99.72 per cent. This homogeneity, except for the traces of phosphorus and sulphur it contains, may explain a mystery which intrigues modern metallurgists: although it has stood in the open for more than fifteen hundred years, it shows no sign of rust.

The Iron Pillar near Delhi in India is a symbol of the toughness of the metal which replaced bronze across the ancient world. It is forged from pure wrought iron, and after more than 1500 years in the open it shows no sign of rust. *Created in honour of Garuda, the representative of the Hindu god Vishnu, during the reign of the Gupta king Chandra II in the fourth century AD, the Iron Pillar stands beside the stone tower of the Qutub Minar in the suburbs of Delhi. It is seven metres tall and weighs six tonnes.*

According to the Sanskrit inscription incised deeply into its surface, the Iron Pillar was created in honour of Garuda, the representative of the Hindu god Vishnu, during the reign of the Gupta king Chandra II in the fourth century AD. It is the largest single piece of forged iron to come down to us from antiquity. When it is considered that it could only have been made by heating and hammering together a great number of billets of iron into a seamless whole, it represents an astonishing luxury of effort and material.

The Iron Pillar is one of the most enduring symbols of a positive, confident commitment across the ancient world to a new metal and a new age: the Age of Iron. It is a mile post of the time by which the making of iron had become a central preoccupation everywhere, and its possession a determinant of success or failure, perhaps even of national survival. For iron had become, by the beginning of the Christian era, something not just to make but to emulate. The name had entered all languages as a synonym for strength and resolution. Before the Iron Age men had tried to be as strong as oak. Now they would seek to develop a will of iron.

The Iron Age changed the way of life of early man in a vast number of ways. It signalled another step away from agriculture and towards industry. It meant more plentiful tools and weapons. Iron axes cleared more forests for more cultivation to support larger populations. And when it came to blood and iron, superior swords and armour won decisive wars. Iron and its alloy, steel, became what it has remained – the most widely used and most useful metal in the world. It is the load-bearing skeleton of our material civilisation.

One great question still hangs over that fundamental advance in technology from bronze to iron: why was it so long delayed? Iron was and still is much more plentiful, and therefore cheaper, than copper. But although it had been known for thousands of years, and even used on a small scale throughout the Age of Bronze, iron did not come into general use until the beginning of the first millennium BC – and only then because of an extraordinary combination of historical circumstances.

We pick up the story in the second millennium BC, with the Age of Bronze at its height. Across the ancient world a variety of metals is in daily use. They include gold, silver, lead and tin, but the bulk of metal production is still in the form of copper and its alloys, especially bronze. Near East metalsmiths have perfected many techniques of metal

By the second millennium BC a variety of metal-working techniques was in use across the ancient world. Those methods have changed little in places such as India. *Melting brass for casting water pots in Ranchi, a small town in the state of Bihar, west of Calcutta.*

Opposite: The high craftsmanship of the second millennium BC is represented by the mask of King Tutankhamun, made of beaten gold, inlaid with lapis lazuli. *The mask, found in the young king's tomb in 1926, is now in the National Museum, Cairo.*

working, particularly those involving the use of the hammer. As mentioned in the previous chapter, this line of development is in direct contrast to the casting tradition in the Far East. It is tempting to see in Western technology a direct descent from the skills of the Neolithic flint knapper.

That high craftsmanship in metal working of the second millennium BC was vividly resurrected in 1926 with the unearthing of the splendours of King Tutankhamun's tomb at Luxor, in Egypt. The young king's body was encased in a series of three coffins, each made of gold. They were not cast, but assembled from solid plates of metal, wrought by hammering. The epitomy of the goldsmith's art is Tutankhamun's mask. The smooth, gleaming surface bears no trace of the thousands of hammer blows which shaped its gentle likeness.

There is one object from the king's tomb, however, which stands apart from the magnificent profusion of jewel-encrusted gold objects now on display in the Cairo Museum. It is a dagger, found beside the young king's body, with a gold hilt and a blade not of gold, silver or copper, but of shining, virtually untarnished iron. No analysis of the blade has ever been permitted, so we have no precise knowledge of its composition or of how Egyptian craftsmen, more than three thousand years ago, prepared this virtually immutable surface. Iron was a metal which at that time was hardly used. For good reason, it was considered inferior to other metals.

Iron had been known, in its meteoric form, for as long as man had been familiar with native metals. But unlike the other and more commonly used native metals, gold and copper, iron was of little interest to early men because they could not melt or cast it. Pure iron does not

More remarkable, in its technology, than any of the splendours recovered from Tutankhamun's tomb is this dagger. The hilt and scabbard are of gold, but the blade is of untarnished iron. *The methods which have prevented it from rusting remain a secret of the Egyptian craftsmen.*

54

melt below 1537°C, and that temperature was quite unobtainable, even when the smeltermen learned to use bellows to blow their furnaces. Therefore the fragments of meteoric iron found lying on the ground could not easily be combined into larger pieces. And since the only technique available for working with iron was hammering, these finds were useful only for small artefacts such as arrowheads or decorative trinkets.

It is not known precisely when iron smelted from terrestrial ores became available, but there is good reason for thinking that its discovery was accidental, and probably occurred first during the smelting of copper. Convincing evidence for this proposition has been provided by the series of excavations by Beno Rothenberg and his teams in the Timna 'copper belt' in Israel.

One of Rothenberg's most important discoveries was a temple dedicated to the Egyptian goddess Hathor, dated to the period during the second millennium BC when the Egyptians were operating a large-scale mining and smelting industry in the Sinai. The temple was built at the foot of a cliff, near the main smelting workshops. In and around the temple were found several hundred small metal objects. Most were of copper, but there were also many small items of worked iron, mainly rings and bracelets. A selection of these has been analysed by the British Museum Research Laboratory with interesting results.[21]

All the copper items were found to contain quite significant percentages of iron – more than would be expected from the copper ore around Timna. When the iron pieces were analysed they were found to contain up to three per cent copper. The iron ore found in the Timna area contains nothing like this percentage of copper. Experiments in Britain at the Institute for Archaeo-Metallurgical Studies provided the answer: the iron was a by-product of the copper smelting.

In the first chapter we saw how the smelter workers at Timna had discovered the value of iron oxide in smelting the local copper ore, malachite, which contained silica sand. The iron combined with the silica to form slag, thus freeing the copper. However, it appears that in the process some of the iron in the hottest part of the furnace was occasionally reduced, along with the copper. It was these bits of iron, removed by re-melting the copper and skimming the surface, which had been used to make the iron jewellery found in the temple of Hathor.

The purification of copper which had been fluxed with iron seems to have been a widespread practice in the ancient world. It had obviously been found that accidentally smelted iron did not blend smoothly with the smelted copper, as did arsenic, lead or tin, but formed stringy lumps. These seriously affected the casting and working qualities of the copper, and therefore as much as possible of the iron was removed. Such material was no doubt experimented with, and perhaps recognised as bearing a close relationship to meteoric iron. The next step would be to try smelting iron oxide separately. But however successful such attempts may have been, they still failed to create a serious demand for iron.

The problem lay with the metal itself. Its melting point of 1537°C could not be reached in Bronze Age furnaces even with forced draught, so that iron, unlike copper, never became liquid and separated from the slag. At the temperatures which *were* achieved, iron could be reduced from its ores, but only to a spongy mass mixed with slag, called a 'bloom'. This bloom became the raw material of the blacksmith. By repeated hammering he could drive out the slag, until he produced a bar of almost pure iron. This came to be known as wrought iron.

Bloomery iron was, however, inferior to bronze in several ways. It was softer, and while it could be hardened to some degree by hammering, under use it still would not hold an edge as well as bronze. Finally, it rusted with alarming speed on exposure to damp air. Bronze, on the other hand, developed a surface film which shut out the air and prevented further deterioration. A comparison between iron tools from Palestine, dated to about 500 BC, and bronze weapons some fifteen hundred years older, from Iran, is instructive. The iron objects have corroded to the point where they are barely recognisable, while the bronze artefacts, cleaned of their patina, are as shining and flawless as when they were made. But events were about to confer on iron a new and enduring role in the march of civilisation.

In the first centuries of the second millennium BC a kind of cosmopolitanism had grown steadily in the Near East and eastern Mediterranean basin. People travelled as never before from place to place across the ancient world. The conduct of human affairs everywhere was transformed by new inventions, from alphabetic writing to the sword and the chariot. There was a flourishing trade in metals. Copper was mined in Anatolia and Cyprus, in large quantities, and supplied to Mesopotamia, Egypt and the new Minoan kingdom on Crete. In these busy centres, large numbers of artisans and craftsmen turned out a never-ending stream of bronze artefacts of all kinds.

Then, around 1200 BC, this animated scene was thrown into chaos. The eastern Mediterranean basin was suddenly invaded by the 'sea peoples'. This vague title is the best that history can do to identify the waves of fierce invaders who swept down from the direction of the Balkans into Greece and Anatolia and on across the Mediterranean, overrunning Crete and Cyprus, to the coast of the Levant, conquering and destroying as they went, plundering the great ports of Byblos and Arwad, until finally they were turned back by the Egyptians. After this defeat some of the 'sea peoples' settled down in Palestine, where they became known as the Philistines.

The turmoil that engulfed the whole of the civilised West saw the annihilation of the great cultural and commercial centres. Established diplomatic and political contacts were broken. More significantly, the trade routes across the ancient world were overrun and cut. This last disruption may have been the key factor in the collapse of the Bronze Age, because of one crucial consequence: the breakdown in tin supplies. Without tin there could be no bronze. An age that had survived for two thousand years was strangled in two hundred, and it never recovered.

The interruption of the tin supplies and its likely consequences has been exhaustively analysed by many authorities, including Jane

Waldbaum, Associate Professor of Classical Arts and Archaeology at the University of Wisconsin. Dr Waldbaum states: 'It is tempting to see in these events and their aftermath a situation in which the main supply link to the eastern source of tin was cut and established trading systems completely disrupted – systems that had prevailed in the Near East and eastern Mediterranean through much of the second millennium. With unreliable or reduced access to an important raw material such as tin, the peoples of the eastern Mediterranean had little choice but to turn to the material nearest to hand and make what use they could of it – and that material was iron.'[22]

This theory, unlike some speculative views of metallurgical history, can cite two well-established lines of material evidence. One is the quite rapid disappearance from the archaeological record of artefacts made of bronze, first of what might be called luxury goods, such as figurines, ornaments, toilet articles and vessels, and then tools, utensils and weapons. In those articles still being made there was an increasing use of melted-down bronze, rather than new metal.

'At the same time,' states Waldbaum, 'there is one phenomenon that occurs in nearly every area equally, and that is the appearance of iron with increasing regularity for the more mundane requirements of existence . . . small quantities of iron are used for tools and weapons nearly everywhere in the twelfth century BC; they increase significantly nearly everywhere in the eleventh; and may be said to be in common, though not exclusive, use nearly everywhere by the tenth. The pattern is remarkably consistent, given the varying nature and quantity of the evidence from each area, suggesting an underlying common reason (or reasons) for the adoption of iron in all these related areas at this time in history.'

But if iron was so inferior to bronze, how was the ancient world able to turn to it so widely and so emphatically, to such a degree that within a few centuries iron had become an almost universal replacement for bronze in virtually every aspect of daily life? The answer lies in the discoveries made by blacksmiths in the treatment of iron which transformed its very character. Where this technological revolution began we cannot be sure, but throughout recorded history one group of people has been persistently credited with the invention of 'good iron': the Hittites of Anatolia. Their appearance upon the stage of history was as remarkable as any in the cavalcade of empires and civilisations.

In the second millennium BC a colony of traders from the city of Assur, on the Tigris in upper Mesopotamia, established themselves in Anatolia. They set up a trading post or *karum* among the bizarre rock formations of Cappadocia, and exchanged expensive fabrics for the copper and other minerals that were so plentiful in Anatolia. These Assyrian merchants also traded tin, the source of which is unknown, through Anatolia to the Mediterranean.

We know a great deal about how the Assyrians operated, because they meticulously recorded their transactions on clay tablets in cuneiform writing. Thousands of tablets were found during excavations of their trading centre at Kanesh. Around 1900 BC unusual names begin to appear on these tablets – names which research has shown to belong

A deity of the Hittites, the mysterious people who emerged from obscurity to establish an empire in Anatolia, and reputedly discovered the secret of making 'good iron' before vanishing once more. *Found on the walls of the ancient Hittite capital of Hattusa during excavations in 1907, this stone relief is now in the Museum of Anatolian Civilisations, Ankara.*

to the Hittites, previously known mostly through vague references in the Old Testament.

Little was known about the Hittites' material culture until about a century ago. Since then archaeologists have unearthed impressive Hittite ruins in Anatolia and northern Syria, and scholars have finally deciphered their language, extinct now for more than three thousand years, but preserved in cuneiform writing on clay tablets. Stone reliefs provide a remarkably life-like impression of their appearance and dress. With their fierce expressions, tall hats, turned-up boots and short tunics they remain one of the most distinctive but enigmatic peoples of the ancient world.[23]

The ruins of Hattusa, the Hittite capital, are slowly emerging from excavations on the uplands of northern Anatolia, over an area of some 160 hectares. The Great Palace includes not only large halls and living quarters, but rows of store rooms, some containing huge storage jars for oil or wine. *The site of Hattusa is near the modern village of Bogazkoy, 200 km north-east of Ankara.*

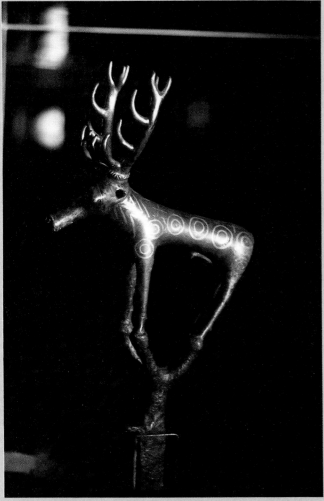

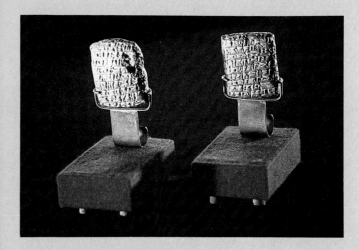

Left: The Hittites drew upon the earlier, already highly developed metal-working technology of the Hattic people. This bronze and gold stag, found in the Hattic royal tombs at Alaca Huyuk, near Hattusa, dates from before 2000 BC. *Museum of Anatolian Civilisations, Ankara. Above*: The Hittite language has been deciphered from cuneiform writing on clay tablets. One such letter from a Hittite ruler to a king of Assyria refers to the 'good iron' for which the Hittites were renowned. *Mining Institute, Anakara.*

Where the Hittites came from before settling in Anatolia has never been established. Their language bears no relation to the Semitic languages of the Near East, but is more akin to the family of European tongues. So far, the only certainty about the land of their origin is that it was somewhere to the north of Anatolia, in that great arc of territories extending around the Black Sea from the Balkans to the Caucasus.

The Hittites established their capital on the windswept uplands of northern Anatolia, and named it Hattusa. It was unlike any other known city of the ancient world. Set on a high ridge, it was surrounded by a massive stone wall nearly six kilometres in circumference. At strategic points the wall was pierced by gateways framed with huge blocks of stone, guarded by stone gods and lions, and sealed at night by heavy doors of timber and bronze.

The Hittite capital covered an area of some one hundred and sixty hectares, and was dominated by palaces, temples, markets, and other public buildings, all expertly built of stone. The Great Palace included not only halls of worship, courtyards, and living quarters, but rows of storerooms for oil, grain, wine and other supplies. Some of the clay storage jars, nearly as tall as a man, still stand where they were unearthed during the decades of excavations which are slowly exposing the city.

From this great stronghold the Hittites marched down across much of the known world. Their armies conquered Assyria and captured Babylon. They went on to dispute control of Syria with that other great expansionist power of the time: Egypt under the Pharaohs of the New Kingdom. The walls of the temple of Ramses II at Luxor record the climactic battle in this conflict of cultures, which took place at Kadesh in Syria, around 1300 BC. It involved more than forty thousand men, and was one of the great battles of antiquity. The Hittites made good use of their speedy two-wheeled chariots and superior weapons, but the Egyptians, rallied by the young Pharaoh, held on, and eventually threw the Hittite forces back. The battle of Kadesh ended in a draw, and led to a treaty of peace between Egypt and the Hittite kings that lasted until 1200 BC.

But then, as elsewhere in the eastern Mediterranean, a tidal wave of fierce European tribes swept into Anatolia, and the Hittites vanished into the oblivion from which they had appeared. So completely was their empire and their culture destroyed that until quite recently their very existence had been difficult to substantiate. Among the mysteries that have still to be unravelled is the question of whether the Hittites really did have the lead in iron-making technology that legend has attributed to them.

That reputation for iron making rests more on literary than physical evidence, but there is clearly a basis for it. For example, among the thousands of inscribed clay letters found among the ruins of Hattusa is one written in the thirteenth century BC from a Hittite ruler to another, thought to be a king of Assyria. 'As for the good iron which you wrote about to me,' runs the letter, 'good iron is not available in my seal house in Kizzuwatna. That it is a bad time for producing iron I have written. They will produce good iron, but as yet they will not have finished. When they have finished I shall send it to you. Today, now, I am despatching an iron dagger blade to you.'

According to the ancient texts, much of the success of the Hittites in their expansion into the Levant and Mesopotamia was due to their possession of superior weapons – swords, spears and arrows – made of 'good iron', which smashed through the bronze shields of their enemies. This advantage was reputed to be derived from the existence within the Hittite kingdom of a subject tribe of skilled metal workers called the Chalybes, who lived on the Black Sea coast in what later came to be called Armenia. It was the Chalybes who had, it was said, discovered the secret of making iron harder than hammered bronze, and for about two hundred years, from 1400 to 1200 BC, the Hittites kept that secret to themselves.

The physical evidence for this supposition is sparse. Certainly, geologists have established the existence of sands rich in iron minerals – magnetite and olivine along the southern coast of the Black Sea. These particular ores are 'self-fluxing', and the iron in them is reduced to metal at the easily attainable temperature of 900°C. They may have yielded a superior form of bloom iron.

The Hittites also occupied, in Anatolia, an area with a long history of metal working. The Hattic civilisation which preceded the Hittites in the third millennium BC produced objects in gold, bronze and silver of outstanding beauty. Among the treasures found in the Hattic royal tombs at Alaca Huyuk, near Hattusa, there was a remarkable dagger with a gold hilt and an iron blade. It has been dated to about 2500 BC, and is one of the earliest known artefacts of man-made rather than meteoric iron.

From all these references and accounts, and the small amount of physical evidence available, archaeo-metallurgists are now prepared to acknowledge that during the period of Hittite rule there were very likely some smiths in that kingdom who had more than a superficial knowledge of iron making. Different characteristics of iron were recognised, and iron with certain qualities could be produced, although not always, apparently, with certainty. Ironsmiths as well as coppersmiths are known to have been attached to the Hittite court, and there was obviously a more than ceremonial interest in the production of iron.

Then came the invasion of Anatolia by the European tribes in 1200 BC, and the downfall of the Hittite kingdom, which by this time included Syria, northern Palestine, and most of Mesopotamia. One effect of this may well have been the dispersal not only of many of the Hittites' subjects, but also of their presumed skills in iron making. Certainly it is true to say that the use of iron in the ancient world generally was on a very small scale until the breakdown of the Hittite empire. After that iron very quickly began to appear everywhere.

For these reasons, therefore, in addition to the textural evidence, we may conclude that it was almost certainly in Anatolia, towards the end of the second millennium BC, that blacksmiths began to make the technical discoveries about working iron which were to transform it into the most useful and most widely used of all metals. The first and most important discovery may well have been made accidentally.

During the process of hammering iron bloom to drive out the slag and reduce it to wrought iron, the blacksmith kept re-heating the bloom in his charcoal furnace. He had to heat it to about 1200°C to make it soft enough to work, and probably never let it go below about 800°C until the job was finished. During this time the iron was in frequent contact with white-hot charcoal, and the carbon monoxide gas produced by the fire. We now know that in such circumstances a small amount of carbon from the charcoal and from the carbon monoxide will diffuse into the surface of the iron. This alloy of iron and carbon is much harder than pure iron. In fact it is steel.

The effect of 'steeling' on the hardness of iron is dramatic, especially by comparison with bronze. Pure iron is much softer than unworked bronze, but the addition of even 0.3 per cent of carbon makes the steeled iron as hard or harder than bronze. Raising the carbon content to 1.2 per cent makes the iron even harder than cold-worked bronze. And if the blacksmith then cold-hammers the steeled iron, it develops twice the strength of cold-worked bronze.

The first smiths to observe this phenomenon may not have understood the physical changes in their material but, depending upon their skill, they might have been able to reproduce it to suit the function of the object they were producing. An obvious case is the steeling of the edge of a blade, rather than attempting to steel the full width of it. Another application of steeling is to harden the tip of an iron pick or chisel. The earliest examples of these forms of selective steeling or carburising are from about 1200 BC.

Having made this discovery, the early smiths seemed to have stumbled upon a second technique which improved the quality of their steeled iron still further. This was 'quenching', the sudden cooling of hot metal by plunging it into water. It is not known when this technique was first used, but there is an interesting comment on it in a paper entitled 'How the Iron Age Began', by James D. Muhly and two archaeo-metallurgists, Robert Maddin and Tamara S. Wheeler:

> One item of literary evidence clearly indicates that blacksmiths of the eastern Mediterranean were familiar with the process in the eighth or seventh century BC. The passage is in the ninth book of the *Odyssey*. Trapped in the cave of Polyphemus, the one-eyed giant, Odysseus and his men manage to get the giant drunk. They decide to blind him by snatching a burning olive trunk out of a fire and thrusting it into his eye. Our translation follows Richmond Lattimore's: 'As when a man who works as a blacksmith plunges into cold water a great axe or adze which hisses aloud, doctoring it, since this is the way that steel is made strong, even so Cyclops' eye sizzled about the beam of the olive.' The description could only have been written by someone who had watched a blacksmith quench the hot iron and knew that the quenching was done to increase the hardness of the metal.[24]

Finally, the blacksmiths of antiquity made a third important discovery. This developed from quenching, although somewhat later, during the early centuries of the Roman period, perhaps in the fourth century BC. Quenching made steeled iron harder, but it also made it brittle. The process sometimes left cracks along the edge of blades, and there were many breakages of knives, swords, and tools. However, if quenched steel is re-heated to about 700°C for a short period, and then allowed to cool, it gives up some hardness but also loses its brittleness.

It was this process, now called tempering, that finally produced the sharp but strong and supple blades which came to epitomise the quality of steel. These three iron-working techniques transformed the properties and the potential of iron. With their widespread diffusion the Iron Age got under way. Bronze never regained its dominance, even when the trade routes were opened up again and tin once more became available.

As already suggested, the steeling of iron seems to have begun in a core area somewhere in the Anatolia-Mesopotamia region of the Near East during the period 1500 to 1000 BC, and then spread across Europe, Africa and into Asia during the next five hundred years. One direction that the Iron Age travelled was established by an unexpected discovery in central Europe barely a hundred years ago.

The picturesque Austrian village of Hallstatt, on a lake between steep mountains, conceals a history that goes back to the very roots of European culture, and which has its echo even today in the folklore of peoples as widely separated as the Danes, the Irish, and the Bretons.

From about 1200 BC, when blacksmiths began making the discoveries about treating iron which made it superior to bronze, it rapidly became what it has remained ever since: the world's most useful and most used metal. *Itinerant iron workers of Rajasthan, India, sell in the market.*

63

The clues to this ancient tradition were found in the Salzkammergut, the steep-sided valley behind Hallstatt.

On the map of this part of Austria the German word for salt is sprinkled through the valleys as a prefix to the place names, as with Salzburg. The reason is that some of the mountains, including those flanking the Salzkammergut, contain huge domes of salt. And salt is a mineral which has had value throughout human history. In many parts of the world it has been used as currency. The Romans considered salt so important to the efficiency of their armies that the legions were paid in it. The soldiers were given a special ration of salt, or the means of buying it – the *salarium argentum*, or salt money, from which comes the modern derivative, salary.

From about 1000 BC the Salzkammergut and its immense resources of salt became a major centre of distribution and exchange. Its inhabitants mined the mountains and traded the salt south-east through the Balkans to the empires of the ancient world around the Black Sea, the Aegean, and the Mediterranean. The significance of this trade, however, was not fully realised until the latter part of the last century, when a remarkable find was made on the side of the Salzkammergut, not far from the present salt mine.

The chance discovery of a skeleton led to the eventual excavation of a large burial ground on the hillside. No fewer than two thousand graves were found, apparently of very early date. They contained a rich variety of grave goods, including pottery, gold objects, and bronze weapons, tools, utensils and ornaments made with great skill. These were typical of the so-called Urnfield culture of the Late Bronze Age, whose people spread their knowledge of bronze working across Europe.

But among the bronze artefacts was, unexpectedly, an assortment of iron swords, battle-axes, bracelets and rings. The dating of the graves to a period extending from about 700 to 500 BC caused a major revision of Iron Age history. No iron of such early date had ever been found in Europe north of the Alps. The discovery pushed back significantly the date of entry of the Iron Age into Europe. And because of the importance of these finds in opening a window into a previously almost unknown period of European history, the name 'Hallstatt' has been applied to the culture that flourished here in the steep valleys of Austria, setting off a chain reaction which was eventually felt on the most remote shores of the Atlantic, the North Sea and the Baltic.

Nourished by the trade between Europe and the Mediterranean civilisations – with metals, salt, and amber going south, in exchange for gemstones, jewellery, wine and oil – and enriched by the developing skills of its metal-working craftsmen, this part of Austria, known in those times as *Noricum*, became a kind of Sheffield of antiquity. The trade routes, the classical channels for the exchange of ideas as well as goods, brought into the green valleys of the Alps the fires and forges of the Near East. From here, in the final centuries of the first millennium BC, the new technologies of the Iron Age fanned out to put all of western Europe on the anvil. They hammered their way to Spain and France, Britain and Germany, Denmark and Norway, in the hands of an

One of the major routes of entry of the Iron Age into Europe was through the valleys of the Austrian Alps, following the trade in salt between places like Hallstatt and the eastern Mediterranean. From Hallstatt the technology was carried right across Europe by the Celts. *Hallstatt and its historic salt mine lie some 70 km south-east of Salzburg.*

emotional, energetic, creative people, who in the following seven hundred years would lay the political and economic foundations of European civilisation: the Celts.[25]

The Celts were a group of related tribes, linked by a common culture, language, and religion, and although they never managed to unite to form an empire, they held sway, at the height of their influence, from Spain to the Black Sea, from the Mediterranean to the Baltic. They were both practical and artistic. They were the first to fit shoes to horses, and

65

iron rims to their chariot wheels (whose span of four feet eight and a half inches was to become the standard railway gauge in Britain two thousand years later). They gave the familiar form to the chisels, files, saws, and other hand tools that we use today. They pioneered the use of chain armour, the iron ploughshare and the flour mill.

The Celts were also lovers of the good life. From the Mediterranean they introduced into Europe the habit of wine drinking and the fashions of personal ornament and fine costumery. They enjoyed music and poetry. From their romantic visions emerged the legends of King Arthur and the Holy Grail, and Tristan and Isolde. They maintained specialist workers in all the crafts: weaving, pottery, glass making, and above all metal working.

From burial mounds, peat bogs, and river beds right across Europe there has emerged over the past few centuries a steady stream of Celtic masterpieces in gold, bronze, silver and iron. They range from magnificent ceremonial cauldrons to sacred collars or torcs, from richly decorated two-handed long-swords to exquisitely wrought filigree brooches and clasps, the forerunners of the safety-pin. Celtic metalsmiths borrowed elements of design from classical and Near Eastern art, and with their own imagination and craftsmanship produced a style of abstract decoration which has lost none of its originality or impact over two thousand years.

But excavations of more humble dwellings and settlements provide a different picture. They show that the impact of the new metal, iron, on the lives of ordinary people in western Europe in the first few centuries of the Iron Age was neither immediate nor dramatic. Tools, weapons or ornaments made of iron, or any other metal, were quite rare in most communities. The point is graphically made at the

66

Historical-Archaeological Research Centre which has been created at Lejre, near Copenhagen, in the form of a prehistoric village.

This Iron Age settlement has been built on an open hillside, between a pond and a forest, to specifications derived from finds elsewhere in Denmark. It is surrounded by a palisade of sharpened stakes, and consists of half a dozen thatched wattle-and-daub huts which are shared by the people and their animals. Students of history and archaeology spend each summer at the settlement, living exactly as research has shown their forebears did. They wear homespun clothes, throw and fire their cooking pots, make their tools and hunting weapons, and provide their own food.

There is a conspicuous absence of metals here. After the collapse of the Bronze Age, that once universal alloy of copper and tin must have become too scarce and costly for subsistence farmers like these. But iron was just as scarce, and was reserved for highly specialised uses, such as the tips of hunting spears. Although a kind of iron ore could be found in bogs in Denmark, it took a daunting amount of labour to smelt iron from it. Experiments at Lejre have shown that the bog ore must first be heated for hours to drive out excess water. It then takes one hundred and thirty-two kilograms of ore and one hundred and fifty kilograms of hardwood charcoal to smelt about fifteen kilograms of bloom iron, in a shaft furnace which is not easy to construct, and which can generally be used only once.

The final constraint on the use of iron in this period was the long drudgery at the forge that was needed to hammer the spongy iron bloom into any kind of useful tool or weapon. This was the only technique available to the smiths. The transition from bronze to iron had brought with it one crippling handicap: because of the unattainably high melting point of iron, the metal workers were quite unable to cast iron the way their predecessors had cast bronze. This failure was to tie them to their anvils and frustrate the mass production of iron goods for another thousand years.

Unknown to the Europeans, however, that particular frontier of technology had already been crossed, far away on the other side of the world, in a country then completely unknown to the West.

CHAPTER 4
Swords and ploughshares

Almost down to our own time the wall, standing for security and deterrence, has been one of the most familiar metaphors in military history. The greatest expression of this symbolism is the Great Wall of China, which snakes across the mountains and valleys of northern China for three thousand kilometres. The Great Wall was begun about 300 BC as a series of isolated ramparts for defence against raiding nomads. It was completed by the Emperor Ch'in, who had overrun and amalgamated six feudal kingdoms. In 221 BC he founded the first dynasty to rule a unified China.

History has repeatedly proved, however, that walls are only as effective as their defenders, and recent archaeological discoveries have identified the critical factor in the historic success of the Great Wall. It was not just its impregnable rock face, its enormous length, or its steep and forbidding approaches. It was the new weapon carried by the defenders on top of the wall which enabled them to hold off and intimidate their attackers. The ghosts of this guardian army, and the secret of their superiority, have now been brought to life by one of the most remarkable excavations of the century.

The story began in 1974, when a party of well diggers found a single clay figure of a soldier in ancient dress buried in a field near Xi'an, the ancient seat of power of the Ch'in Emperor. This led eventually to the unearthing of the most astonishing deathwatch ever mounted over a departed mortal: the now famous army of life-sized men and horses, made of terracotta, standing guard in front of the hill which contains the tomb, as yet unopened, of the Emperor.[26]

The figures stand in closely ordered ranks in three shallow underground chambers, which were originally roofed with timber and covered with earth. At some point the timbers were burned, and the earth fell in, burying the figures and effectively protecting them against further destruction. Chinese archaeologists have determined the size of the chambers by test excavations, and are steadily exposing the

Above: Recent archaeological discoveries in China have revealed the secret weapon which made the Great Wall so effective a deterrent against the barbarians during the reign of the Emperor Ch'in. *This section of the Great Wall, partially restored, is a two-hour drive north-west of Peking. Left*: Following a chance discovery in 1974 of a single buried figure, a remarkable terracotta army is being excavated from the fields in front of the tomb of the Emperor Ch'in near Xi'an. *Chinese archaeologists estimate that at least 6000 men and horses stand in formation in this one site, which may take decades to excavate fully.*

69

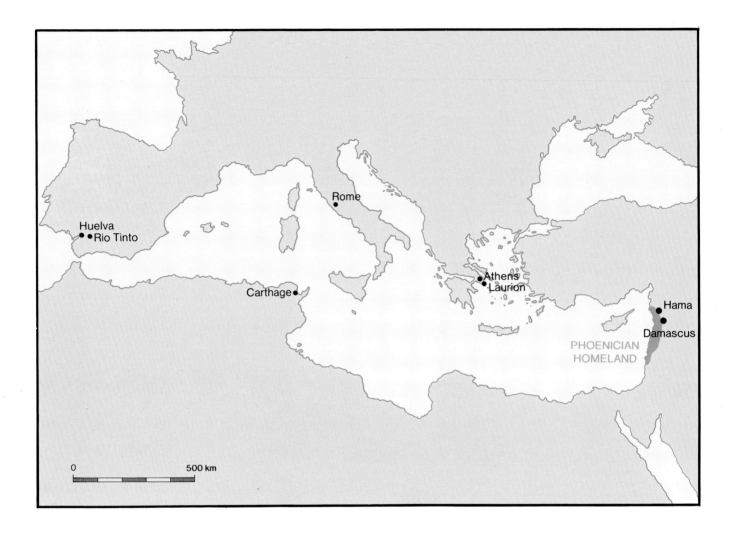

soldiers, horses, and chariots which stand there in formation. It is estimated that the largest chamber contains no less than six thousand men and a small number of horses. This is now covered by a huge hangar-like building, to preserve the site. A second chamber is filled with phalanxes of kneeling bowmen, ranks of cavalrymen and horses, and chariots. The third and smallest chamber is occupied by a command unit of officers.

Many of the figures are crushed and broken, but as they are resurrected and restored on the spot where they once stood, an extraordinary first impression is being confirmed. Although the bodies and limbs were made to a number of stylised patterns, no two faces are exactly alike. The archaeologists believe them to be likenesses of the Emperor's actual soldiers, copied in clay to guard him in death as they did in life.

In this make-believe army, frozen in time for more than two thousand years, one detail is real – the weapons. A whole armoury of swords, spears and short arrows has been found, all made of solid metal. When uncovered they were coated with a heavy patina, like bronze, but cleaning has revealed them to be in pristine condition, their coppery surfaces smooth and shining, the sword edges still keen and the triangular arrowheads still sharp.

70

The Emperor's clay warriors were armed with real weapons of sophisticated metal alloys, including these arrows. *Analysis has shown the alloys to be chiefly of copper, lead and tin, but with significant additions of such 'modern' metals as chromium, cobalt and titanium.*

Analysis of the materials has provided the explanation for their remarkable state of preservation: the weapons are made of alloys of up to fifteen different metals. The major ingredients are the classic components of bronze – copper and tin – but there are others which have only come into use in the West in recent times. One sword contained copper, tin, lead, iron, and measurable percentages of zinc, titanium, molybdenum, nickel, vanadium, cobalt, magnesium, silicon, aluminium, manganese and niobium. The presence of some of these exotic metals might be explained by natural contamination of the copper ores used, but most of them must have been deliberately introduced. The metallurgy of the weapons indicates a Chinese command of alloying, in the first millennium BC, that was quite unknown in the West.

Another kind of innovative technology was represented by unusual assemblies of interlocking metal pieces, found in association with some of the kneeling soldiers. When one was taken apart it was identified as the trigger release mechanism of a crossbow, the wooden body of which had decayed and disappeared. The mechanism consisted of five quite intricately-shaped components, cast in bronze, which were fitted together and mounted at the rear of the wooden 'barrel' of the crossbow. This was an open channel to guide the projectile.

To load the weapon, the string was pulled back and held by a metal finger, and the short bronze arrow with its heavy triangular head was placed in the channel. The weapon was held to the shoulder like a gun, and discharged by pulling the trigger. At short range it was lethal, for the projectile would penetrate any bodily covering, even armour, and the bow could be loaded and fired with great rapidity. It was this device which enabled the Emperor Ch'in's soldiers to hold the barbarians at bay along the Great Wall. The crossbow, invented in China,

71

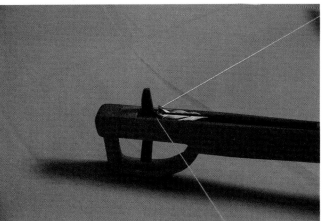

was a weapon that would not be seen in the West for another six hundred years.

Sophisticated alloys and weaponry were not the only important advances in metallurgy taking place in China in the first millennium BC. In that same period the Chinese were mastering a technique of metal working with enormous potential, the casting of iron.

The invention and exploitation of cast iron by the Chinese, some two thousand years before it was used in the West, must rank as one of the most extraordinary chapters in the whole history of metallurgy. It is a story which until quite recently was little known and in fact hardly credited outside China. The material evidence for it has only been found in the last two decades, and much of this has yet to be published in the West. Indeed some of the most impressive finds were filmed for the first time for our own television series.[27]

As already argued, it seems highly likely that bronze was developed independently in China. The same cannot be said for iron. There is evidence of the use of meteoric iron in the second millennium BC, but no man-made iron is known from before the seventh century BC. This fits established Western theory, which holds that iron making reached China as part of the great diffusion of iron-making technology across the ancient world from the Near East, beginning around 1000 BC. The techniques which reached China, as elsewhere, would have been confined to the skills of the blacksmith: the production of bloom iron, its conversion into wrought iron by hammering, and the steeling of wrought iron by carburising in the forge, quenching and tempering.

When iron was taken up in China, however, it was launched on an entirely new trajectory. In the hands of the descendants of the Shang bronze casters, iron was treated in a different way. With their superior furnace technology the Chinese were able to do with iron what they had done with bronze: they melted it. They made iron run like water, and they cast it the way they had cast bronze. It was an achievement which would not be matched in the West until the Industrial Revolution began to gather momentum in Europe.

That period in Chinese history from about 700 until 206 BC, when the Han dynasty succeeded that of the Emperor Ch'in, is still only mistily glimpsed through somewhat vague written accounts. But in recent years a steadily growing body of archaeological evidence has enabled us to identify at least some of the key stages in the development of iron casting, especially during its most productive period, the four hundred years of Han rule. During this great flowering of innovation the Chinese did not ignore those other techniques of iron working, which they had received from the West. They made good use of wrought and steeled iron, in fact they improved upon them. But very early in their involvement with iron they discovered how to cast it, and this became the main thrust of their iron-working technology for the next thousand years.

It will be remembered that the biggest difficulty in smelting iron is its high melting point, 1537°C. Below this temperature pure iron does not become liquid, and therefore will not separate from the slag. The Chinese were able to overcome this problem by a combination of two factors, one of which was the product of their own command of furnace technology, acquired through their long history of pottery making and bronze casting; the other was a fortunate accident of molecular physics whose effect, if not its cause, they well understood. The first factor was their ability to achieve higher furnace temperatures than were obtainable elsewhere. Part of this success was due to superior furnace design, but to it they added two Chinese inventions, the horizontal bellows and the double-acting box bellows.

Opposite: Among the metal objects found with the Emperor Ch'in's soldiers were sets of five oddly-shaped parts, cast in bronze, and apparently designed to interlock. When assembled they proved to be the trigger release mechanism of a Chinese invention, the cross-bow. To load it, the string was pulled back and held by a metal finger, while a short bronze arrow was placed in the groove. The weapon was held to the shoulder like a gun, and discharged by pulling the trigger. At short range it would have been lethal, for the projectile would easily penetrate any existing armour. It was this device, whose equal would not be seen in the West for another 600 years, which enabled Emperor Ch'in's warriors to defend the Great Wall. *Reconstructed bow, with actual release mechanism, in the museum at the Ch'in excavation near Xi'an.*

The typical hinged bellows in the West, even into modern times, was operated by a handle which the blacksmith pumped up and down. Gravity assisted the downstroke, but resisted the upstroke. The Chinese found a different way: they hung their bellows from an overhead beam, so that the direction of pumping was horizontal. This neutralised the effects of gravity, and enabled the bellows to be made very large. The horizontal push could be delivered by men, animals, or a rod from a waterwheel. A scale model in the National Museum in Peking, reconstructed from ancient illustrations, shows such a bellows suspended from a heavy beam. It consists of several linked chambers made of leather, resembling a concertina, and is compressed by a large swinging horizontal rod, like a battering ram. Such a device is clearly capable of delivering very large volumes of air to a furnace.

The double-acting box bellows was even more ingenious. It consisted of a long box containing two chambers, one above the other. A handle connected two rods which projected from the end of the box. The inner ends of the rods were linked to square pistons in the two chambers. A pipe led from the side of the box to the furnace. When the operator pushed and pulled the handle, simple flap valves directed the air flow between the two chambers and then into the pipe, so that one piston pumped on the forward stroke, and the other on the backward stroke. The result was a virtually continuous flow of air to the furnace.

This bellows is still used, single-handed, by backyard blacksmiths in China to blow small forges, but in Han times, according to tomb paintings, it was scaled up in size until it needed several men to operate it. Such a pump delivered a great volume of air, and permitted the use of very large furnaces and massive charges of fuel and ore. The burning of large quantities of carbon fuel in relation to the amount of iron ore being smelted introduced the second factor mentioned above.

When iron is smelted in an atmosphere with an excess of carbon monoxide, some of the carbon enters the iron and has a very interesting effect: it lowers the melting point of the iron. The iron-carbon alloy becomes liquid and flows freely from the slag at about 1150°C, a temperature which was well within the range of Chinese furnaces. The Chinese had found a way to cast iron.

It should be pointed out that this combination of powerful air blast and a high combustion rate of charcoal was not obtained in Western furnaces until waterwheels were used to drive large bellows in the fifteenth century AD. F. R. Tylecote considers that 'the development and introduction of the blast furnace in Europe is one of the most interesting subjects in the history of ferrous metallurgy. We know that the blast furnace was being used in China long before it was used in Europe, but it cannot be assumed that it did not have an independent origin in Europe. On the other hand, its introduction in Europe came at a time when contact between East and West was well established, and all that was needed in Europe was an appreciation of the usefulness of cast iron.' This last reference is to the fact that although cast iron had sometimes been made in bloomery furnaces in the West by accident, as early as Roman times, it was invariably thrown away because it was brittle, and considered useless.

Among the cast iron articles made during the Han dynasty (202 BC to AD 220) were plough-shares and hoes. *In the examples found at the blast furnace site at Zheng-zhou, and displayed by Michael Charlton, only the hardened cutting edge of the ploughshare remains.*

The development of iron casting in China reached its peak during the Han dynasty, as is shown by the discoveries made in Henan province in central China in 1976. The previous year, people working their fields near the city of Zheng-zhou had found a large hoard of ancient farm implements. The tools, numbering more than four hundred, were chiefly adzes and shovels. When they were examined by archaeologists from the city museum they were found to be made of cast iron, and some of them bore the markings 'Henan No. 1'. A search of the site unearthed pottery and bronze coins which firmly dated the finds, including the iron objects, to the latter part of the Han period, which extended from 206 BC to AD 220.

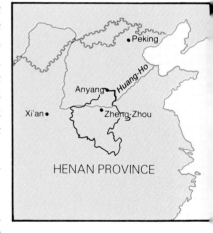

It was not a complete surprise to the Zheng-zhou archaeologists, therefore, when they found, during the spring ploughing of the area in 1976, the remains of two large Han blast furnaces. Iron tools bore the markings 'Henan No. 1'. They had discovered one of the major industrial sites of the Han dynasty, which according to written records had 'nationalised' iron making under forty-nine regional offices in the main producing areas.

Excavations revealed impressive evidence of the scale of these iron works. One furnace still has its oval hearth intact. The hearth, made of rammed heat-resistant clay, and burned grey with use, is four metres long and nearly three metres wide. It rests on a foundation of clay reinforced with pebbles, three metres thick. The furnace wall was of refractory material, a metre thick, reinforced with an earth bank, and is calculated to have been about three metres high. The furnace had a capacity of about fifty cubic metres, and was fed with iron ore, charcoal and limestone flux from the top. Pre-heated air was blown into it through large tuyeres or clay pipes in the wall, twenty-six centimetres in diameter. The molten iron was tapped through a plug near the base of the furnace, and was either run straight into moulds, or cast as 'pigs' for future use. The furnace worked day and night, and production was probably several tonnes of iron a day.

The archaeologists obtained confirmation of their reconstruction of the operations when they dug up several very large lumps of iron near the furnaces. The largest weighs twenty-three tonnes, with a shape corresponding to the bottom of a furnace. It was either left to solidify in a furnace through some interruption of production, or else it collected in the bottom of a furnace as the lining wore away, until eventually the structure had to be pulled down for it to be removed. (Such residual masses of iron, called salamanders, still form in furnaces today.) Since the huge lumps of iron were impossible to cut up and use, they were simply buried. Analysis showed the iron to contain 3.5 per cent carbon, and traces of manganese, sulphur, silicon and phosphorus. It is comparable to high-quality pig iron produced in a modern blast furnace.

One use to which the iron was put was made clear by the finds of scores of both clay and iron moulds for making ploughshares, farm implements, toothed gears, and wheel bearings for carts. More than three hundred of the finished articles were unearthed on the site. The clay moulds, baked hard, were used to make the iron moulds for the actual casting of the implements. Fourteen pottery kilns on the site turned out the clay moulds, as well as the tuyeres and fire-bricks for the furnaces. These finds, and others in neighbouring regions of central China, are an indication of the extraordinary growth of the iron industry in China in the first millennium BC.

The Iron Age in China may have started later than in the West, but the handicap was overcome with dramatic speed. The earliest piece of man-made or smelted iron is a forged dagger, found in Gansu province, and dated to about 650 BC. Soon after, swords forged from smelted iron appear. Then the pace quickens, as the new type of iron comes into use. By 500 BC all kinds of objects are being cast in iron, containing up to 4.5 per cent carbon.

At this point, metallurgists might raise an objection – which may well have been one reason why Western authorities have been reluctant to accept the story of Chinese supremacy in iron casting. The objection is that cast iron with such high percentages of carbon is certainly very hard, but it is also extremely brittle. It does not bend when overstressed, but snaps like a biscuit. How could such a material have possibly been used for agricultural implements, gears and wheel bearings? The answer, as the evidence shows, is that the Chinese had discovered how to transform cast iron into something much more useful: malleable iron, which is similar in most of its properties to wrought iron.

As early as 500 BC, it seems, iron workers had learned that if cast iron is heated to about 800 or 900°C in the presence of air, the oxygen removes some of the carbon from the surface of the iron. As the carbon content falls, the iron loses its brittleness, and a tough 'skin' forms on the cast iron core. When the carbon has been reduced to between one and two per cent, the iron has in fact acquired a steel jacket. It is not unlike the steel the Western blacksmith produced by carburising wrought iron in the forge. So both East and West had come to almost the same result, though by opposite routes.

There was, however, one profound difference. The Chinese could shape their iron articles with ease, by casting, before steeling them, whereas the Western blacksmith could shape and steel his iron only by laborious hammering. And so the Chinese proceeded to turn out steel-jacketed iron tools and implements by the thousand, using moulds in an early form of mass-production. The way they did this has only recently been established, by the discovery of their techniques of 'stack casting'.[28]

In the early 1970s, four large Han iron foundries were excavated in Henan and neighbouring provinces. On the sites were found hundreds of sets of intact, baked clay moulds for casting iron articles. The moulds were arranged in stacks, one on top of the other. All were fed from a single channel down the centre of the stack, into which the molten iron was poured. Some stacks were capable of producing up to one hundred and twenty castings at a time, of such things as cart bearings, harness buckles and bits, and coins. Many of the stacked moulds were in such good condition that archaeologists were able to use them to cast iron articles, identical to those first produced this way nearly two thousand years ago. The first experiments, incidentally, were a failure; the castings were pitted with bubbles and shrank excessively. The investigators succeeded only when they first heated the moulds in a furnace to 600°C before filling them with molten metal, as the ancient iron casters must have done.

These advances in the first millennium BC raised the output and lowered the cost of iron products. They made tools widely available, and thereby accelerated the development of agriculture and the digging of irrigation ditches and canals. By the middle of the Han period the Chinese ironmasters were able to meet the needs of an agricultural nation with a population of eleven million. Then came the most technologically significant advance of all. It gave China an even greater lead over the West, and could have made her the dominant world power — but, for reasons which are still obscure, it did not have that result.

The critical discovery seems to have been made early in the first century AD. Knowledge of it was first obtained from the excavation of yet another large Han furnace site in Henan, containing artefacts marked 'Henan No. 3'. This furnace was different from all others, however. It was designed to turn iron into steel, but not by 'roasting' already cast objects. Instead, carbon-rich cast iron, either straight from a smelting furnace or from 'pigs', was melted in a large crucible. It was then converted directly into steel by 'puddling'. That is, operators stirred the molten iron so that oxygen in the atmosphere drew off carbon from it. By this method the Chinese were able to reduce the carbon content of bulk cast iron to less than one per cent. What they produced was steel, essentially no different from that made in steel plants today by the process invented (or, more accurately, re-invented) by Henry Cort in England in 1784!

The story of China's early mastery of iron and steel production throws into sharp relief a fascinating and puzzling question: why did she not maintain or even increase that advantage? All the conditions seemed to exist for China to lead the world into an industrial and scientific revolution, centuries before it occurred in the West. The products of Chinese technology which reached the West brought about radical changes. They included chemical medicines, metal alloys, paper and printing, the compass, mechanical clockwork, silk weaving, the horizontal loom, the spinning wheel, the waterwheel, the crossbow, and the humble wheelbarrow. The stirrup from China gave the European knight in armour his thrusting power with the lance, and helped to create the feudal system, while Chinese gunpowder was a significant factor in ending it. But these revolutionary advances had less impact in China itself, where stagnation and eventually decadence set in.

Historians now believe that the reasons were both political and social. China's belief that it was a universe in itself led to a dangerous contempt for developments elsewhere. Innovations did not bring about radical social change in the stable, civilised East, as they did in the restless, more dynamic West. While the large and growing population continued to provide cheap energy there was no demand for labour-saving devices. Because most people were poor there were no potential markets to stimulate new industries. China was therefore unable to generate those surges of demand which present challenge to ambitious men and opportunity to the entrepreneur.

One major factor in the decline was the stultifying effect of the most powerful political force in China, the Confucian bureaucracy of civil servants, under the emperor. Jealous of their authority, the bureaucrats stifled new knowledge and ideas, prevented the rise of the individual inventor, and in the end anaesthetised technology. It is a conflict which is still at the centre of contemporary Chinese politics. And so the ironmasters of the Han period, who were well placed to launch a power-driven, mechanised Industrial Revolution, went on making hand-held hoes and ox-drawn ploughs.[29]

Meanwhile, fresh intellectual and practical impulses in the West had begun to raise new powers, as the centre of gravity of the ancient world moved from the Near East to the Mediterranean basin. But before returning to follow the course of the Iron Age through the fortunes of Greece and Rome, there is one curious contribution to the story of metallurgy, originating in the cultural limbo between East and West, which should be mentioned. It is the invention of 'wootz' steel in India.

Wootz was the name given to a particularly tough kind of steel which was in great demand all over the ancient world, from about the middle of the first millennium BC until well into the Middle Ages. It came in only one form: small, round, flat ingots, and it was reputed to make the finest swords and daggers obtainable. The Romans traded gold and silver coins for it, to make their famous short-swords. The Moorish

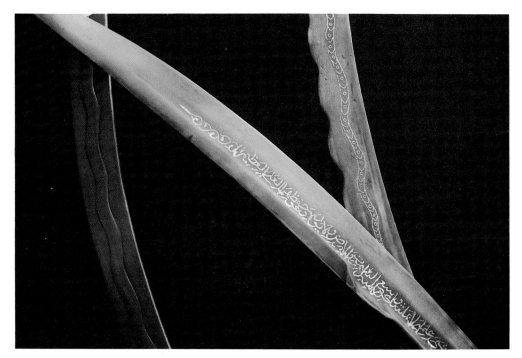

The blades of the famous Damascus or 'Damascene' swords used to carve out the Arab empire in the sixth century AD were made of a particular kind of high-quality steel, originally invented in India, where it was known as 'wootz'. Sometimes, as here, the steel was inlaid with spiritual exhortations in gold. *Military Museum, Damascus, Syria.*

horsemen who carved out the mighty Islamic empire in the sixth century AD did so with their fearsome scimitars of Damascus steel, a term which was first applied to wootz from India, then to a local version of it.

There are two interesting questions about wootz steel: what made it superior to all other steel in the Western world, and how did it come to be invented in a region not otherwise noted for advanced technology? Neither question is easy to answer, since the early history of Indian technology is still largely unknown. However, from the few available sources, including two books recently published in India, it is possible to suggest how, if not why, wootz steel was made.[30]

The first metal-using people in the Indian subcontinent belonged to one of the greatest and yet still least understood civilisations of antiquity – that which flourished comparatively briefly in the valley of the Indus river, which flows through what is now Pakistan. The existence of the Indus civilisation has only been known since 1921, following the discovery and excavation of its twin capital cities, Harappa and Mohenjo-Daro. Its origins in the fourth or third millennium BC are still obscure, but it appears to have reached its peak around 2300 BC. By 1500 BC the Indus civilisation was in rapid decline and soon vanished, apparently after disastrous floods had destroyed its food-producing capacity.

At its height the Indus civilisation was even more extensive than the other two great river civilisations in the valleys of the Nile and the Tigris-Euphrates. Its populous cities and villages were supported by a highly efficient agriculture, external trade and specialised crafts. But the level of culture, from what is known, was not particularly high. The pottery was mass produced and uninspiring. The metal work appears rather backward, compared to what was happening elsewhere, and the copper and bronze tools and weapons show no great skill or quality.

Against this background, it will not seem surprising that the Iron Age began fairly late in India, despite the presence in that sub-continent of some of the richest and most plentiful iron ores to be found anywhere. Iron working seems to have reached India from Iran or Afghanistan around 1000 BC, but the pattern of its adoption is little known. Judging by the results of excavations, iron articles were being widely made and used by about 600 BC, but there has been no analysis which would establish the methods of working or the level of technology.

It is therefore difficult to explain the development which occurred about this time, reputedly in the area of Hyderabad. There, iron workers evolved a quite complicated but highly successful method of making steel, of a quality which was not approached anywhere else. The method went far beyond the technique then in general use for steeling iron by heating it in a forge, so that it absorbed carbon from the burning charcoal. The Indians made small clay crucibles about the size of a rice-bowl, and filled them with short pieces of wrought iron, together with about four to five per cent by weight of wood and the leaves of specific plants. The crucibles were sealed with clay and placed in a pit about a metre deep, which was then filled with charcoal. The fuel was lit, and an air blast from bellows directed into the pit to raise the temperature. The firing continued for several hours.

When the iron inside the sealed crucibles melted, the carbon from the plant material became evenly distributed through the molten liquid. Finally the liquid steel, as it had become, was poured into stone moulds to produce the typical wootz ingot: a flat cake about twelve centimetres in diameter weighing around one kilogram. Because of the uniform distribution of the carbon through the iron – something which had never been possible by forging and hammering, or even by the later method of heating bars of iron with charcoal in a furnace – wootz steel was remarkably homogeneous, and superior to anything else known.

The Indian steelmakers went to great lengths to conceal the secret and even the precise origin of wootz, and sold it to the Mediterranean countries by roundabout routes through centres of trade such as Damascus, and even through Abyssinia. It was not until the seventh century AD that the art of producing steel in crucibles reached the Near East from India. The Arab metalsmiths in Damascus, and later in Toledo in Spain, then under Moorish occupation, learned how to apply the Indian techniques to local ores to make the famous Damascus swords. Cakes of crucible steel, some containing titanium apparently from a mine near Damascus, were blended by forging and working, to produce the shining, flexible blades with the highly prized 'damascened' or water-patterned finish. (After the disintegration of the great Islamic empire in the tenth century, the secret of making crucible or wootz steel seems to have been largely forgotten until, as described in a later chapter, an English clockmaker, Benjamin Huntsman, re-discovered it in the eighteenth century, and devised the type of steel which was to make his home town of Sheffield world famous.)

At the eastern end of the Mediterranean, the latter half of the first

millennium BC saw new powers move on to the world stage as the older cultures of the Near East sank into the desert sands. More than any of their predecessors, the classical civilisations would depend upon a wide range of metals for their trade, prosperity and very survival. There is no better example of this than the glory that was Greece, and the silver mines of Laurion which saved it from premature oblivion.

Of the new city states that grew among the islands and rocky peninsulas of the Aegean, none was to become more famous than Athens. The majestic ruins of the Acropolis are among the best known monuments to that flowering of art, literature and philosophy, which was nourished by the most sublime resources of the human intellect. But Athens also rested on more mundane pedestals, the ruins of which are to be found sixty-five kilometres to the south near Cape Sounion, the bare headland on the coast of Attica. Here, on the hillside overlooking the little village of Thorikos, are the shafts, washing tables and marble storage cisterns of the silver mines of ancient Laurion, which financed

'The glory that was Greece', represented here by the Temple of Poseidon on Cape Sounion, was founded not only upon the great forces of the intellect that flourished in Athens, but upon the flood of silver from the mines in nearby Attica. With this wealth the Athenians built the fleet which defeated the invading Persians at Salamis in 480 BC. *Cape Sounion lies 40 km south of Athens.*

81

the sword of Athenian power and enabled it to dominate the Aegean for the best part of the fifth century BC.[31]

The southern tip of Attica was one of the most famous mining regions of antiquity. The Myceneans or perhaps the Cretans in the second millennium BC may have been the first to dig into the veins of silver-lead in the reddish limestone slopes. However, it was not until the rise of the city state of Athens in the sixth century BC that Laurion assumed its true importance. At first, the Athenians had obtained their wealth from silver and gold mines in Thrace and Macedonia. When these were captured by the Persians in 512 BC, they had to turn their attention to Laurion, even though the surface seams – the 'first contact' – appeared to be worked out.

Despite the part that Greek civilisation was to play in shaping the concepts of individual freedom and social democracy, the mines the Athenians now opened up in Attica were worked by slaves, in conditions of extreme hardship. The contemporary Diodorus Siculus wrote that one form of slavery in Hellenism stood apart from others; the mines were 'a Hell on earth which neither Stoicism nor Delphi could touch, and kings and cities were equally guilty'. Wielding the iron picks and hammers that had by this time replaced bronze, and directed by Athenian overseers, the slaves hacked shafts and tunnels into the rock, following the veins of silver-lead ever deeper.

At the beginning of the fifth century BC, within a decade of reopening the mines at Laurion, the Athenians struck the 'third contact', a fabulously rich layer of high grade silver-lead. Large-scale separating plants were built on the hillsides to treat the ore. Cisterns for water storage, up to seven metres across, were hollowed out of the rock, or made from closely-fitting blocks of marble. Slaves carried large jars of water to the washing tables in marble-walled buildings, where the crushed ore was sluiced across serrated stone slabs to separate the heavier silver-lead from the waste material.

Excavations have exposed the great extent of the operations, and walking across the hillside it is not difficult to hear echoes of the intense activity that once marked this place: the pounding of hammers as the slaves crushed the ore, the bustle of water-bearers to and from the cisterns, the ceaseless splashing from the washing tables, and above all the din the crack of the whips and the shouts of the overseers. From these mines a river of silver poured into the Athenian treasury, and into the pockets of the Athenians, for all citizens shared in the profits from the mines.

However, in 483 BC the Athenians were persuaded to forgo their dividends from Laurion, in the face of a looming threat from Asia. Huge armies under the Persian king, Xerxes, had crossed the Hellespont from Anatolia. They had occupied Thrace and Macedonia, and were advancing remorselessly down through the Greek peninsula. The statesman Themistocles convinced his fellow Athenians that their only hope of survival lay in the construction of a 'wooden wall of ships' to fight the Persian navy. The Greeks used their silver to build an extra one hundred and thirty triremes or fighting galleys to strengthen their small fleet of seventy ships, and waited for the Persians.

In 480 BC the Persians defeated the smaller Greek army at Thermopylae, and soon after captured and destroyed Athens. Greek civilisation was on the brink of extinction. But when the Persian fleet followed the Greek navy into the narrow waters between the island of Salamis and the mainland, the Greek ships suddenly turned in desperation and attacked their vastly superior enemy.

The Battle of Salamis was the greatest naval encounter of the ancient world. The Persian fleet was virtually annihilated by the more agile Greek ships, and Xerxes' dream of conquering the Aegean and the Balkans was shattered. From this victory, and with the renewed flow of silver from Laurion, Greek culture rose to its magnificent climax in the adornments of Athens under Pericles. Among the reminders of his time, and of the debt that Athens owed to the silver mines of Laurion and their legions of slaves, are the Parthenon in Athens and the Temple of Poseidon on Cape Sounion, which looks across at the hillsides with their tumbled marble washing tables and overgrown mine galleries.

By the third century BC the silver mines in the hills of Attica were spent, and so was the energy of the Hellenistic world. Once again the link between political influence and the control of material resources to sustain it was due to be confirmed, in the rise of a new power in the Mediterranean. Inseparable from this new domination would be the output of even more productive mines than Laurion, which would

The vast, coloured spoil heaps of the Rio Tinto mine in Spain are the product of a history of mining for gold, silver and copper which extends back for at least 5000 years. *The name Rio Tinto means Red River, and comes from the stream which rises in the mine area, its waters stained by the iron washed out of the hills.*

provide the military, financial and economic backbone of the world's first technological empire.

The might of Rome, even after nearly fifteen hundred years, is still almost palpable. The indestructible ruins of its monumental achievements still stand, across the vast area of the world that it once commanded. Yet few victories the Romans won were more significant than their conquest of immense mineral resources, from one end of the Pax Romana to the other: the gold of Gaul, Wales, the Balkans, and Persia; the silver of the Pyrenees, Greece, and Anatolia; the copper of Cyprus and the Sinai; and the tin of Cornwall. But from the first century BC one country above all came to provide the mineral wealth of the Roman empire. That country was Spain, and of all the rich mines in that mountainous land none could match the fabulous wealth of Tartessus, as it was in Roman times, or Rio Tinto, as we know it today.

Buried in a labyrinth of vividly coloured and sculptured hills and valleys, some forty kilometres inland from the Gulf of Cadiz in southern Spain, lies one of the greatest belts of minerals on earth. The gigantic ore body contains gold, silver, copper, iron, zinc and other metals. They were originally deposited on the sea-floor by undersea volcanoes, spouting mineralised salts from the earth's interior. First they were buried beneath sediments, which turned to rock, and then the great mass of minerals was eventually uplifted by crustal movements as the seas retreated and the land emerged. In time, the bands of coloured ores were exposed by erosion, and began to stain the waters of the streams that flowed from the valleys cutting through the formation. The largest of them became known as the 'red river', and Rio Tinto it has remained ever since.

Today the visitor, approaching Rio Tinto over the limestone hills dotted with silver-grey olive trees, is suddenly confronted with one of the most bizarre landscapes on earth. It is dominated by the immense mountains of earth and rock, in all colours from red, yellow and brown to green, purple and black, which are the result of thousands of years of mining.

The present operation is a gigantic open cut, which is eating down into the central mountain. All around are reminders of past labours: Roman drives running straight into solid rock, their entrances half obscured with ferns and mosses which seek the water trickling from the darkness; the old village of Rio Tinto, with its collapsing white-washed red-tiled cottages, where the Spanish mine workers lived in the eighteenth century; the abandoned boiler houses and smelters, with the huge brick flue that ran up the hillside like a covered corridor to discharge the fumes, until the smelter was moved elsewhere; endless rusting railway lines and rows of derelict railway engines and trucks, a legacy of the British period of ownership, which began in 1873 and ended in 1954.[32]

No one can be sure when the first men came up the stark valleys of Rio Tinto, attracted by the coloured ores in the hillsides, but there is now firm evidence that mining has been carried on here for at least five thousand years. Recent finds of stone hammers and granite pestles and mortars for grinding up the ores, and shallow trench mines in quartz outcrops, have established beyond doubt that mining began in this part of Spain as early as the third millennium BC. Those Chalcolithic miners were seeking malachite, for the copper it contained. No copper-smelting furnaces have yet been discovered, but slag similar to that found at Timna in the Sinai desert of Israel suggests that a comparable level of technology had been reached here, with smelting being carried out in simple bowl furnaces.

The ability of Chalcolithic man with simple tools to mine in rocky country was limited, so once the surface deposits of malachite were exhausted there was little he could do. During the second millennium BC there appears to have been little activity around Rio Tinto. What happened from then onwards has been made clear by a remarkable accident of recent times. Among the modern workings, archaeologists have found an extraordinary chronology of three thousand years of mining. A road driven through an ancient mining area has exposed in the side of the cutting the layers of mining debris from the past. It is a cake of history. The bands of slag vary from red to almost black, and contain bits of broken pottery which have enabled them to be accurately dated.

At the bottom of the face a shallow depression has been excavated, and dated to the Bronze Age. From the debris found in it, the archaeologists believe that it might have been some kind of rubbish pit used by the miners seeking copper. Above this there is a layer of distinctly different colour. It marks the earliest appearance of silver slag rather than copper, and it coincides with the arrival in Spain of the Phoenicians, around the ninth century BC. These adventurous sailors and traders from Tyre and Sidon ranged the whole length of the Mediterranean, and perhaps even round the west coast of Africa. They were attracted to Spain, it appears, by tales of the rich mines of a people called the Tartessians. The style and dates of the pottery the Phoenicians left in Spain suggests that they virtually monopolised the metal trade for two centuries. They took silver away from Rio Tinto in such quantities, according to one account, that when their ships could load no more they threw away their iron anchors and substituted ones cast in silver.

The next layer of slag and mining rubble was laid down by the Carthaginians. To them the mines of southern Spain were the plunder of victory. The silver that continued to pour from Rio Tinto was the mainstay of the empire they established in the central and western Mediterranean in the fifth century BC. When Hannibal was defeated and Carthage overwhelmed, as the Punic wars ran their course, the Carthaginians gave way to the great power that was to control the ancient world from the Atlantic to the Euphrates, from the North Sea to the Sahara – the immensity of the Roman empire.

The great bulk of the smelting slag beside the road cutting at Rio Tinto, perhaps six metres in depth, was produced during the five or six hundred years that the Romans operated the mines they called Tartessus, beginning about 200 BC. Over the whole area of the workings, archaeologists have identified Roman slag heaps containing more than ten million tonnes of smelted material.

The first two centuries of Roman occupation of Tartessus, until the beginning of the Christian era, were under the Republic, and the scale of operations was not very different from Carthaginian times. Silver was still the chief product. The picture changed dramatically in the first century AD, with the beginning of the Roman Empire, under Augustus. To meet the demands of their growing economy, and the appetites of the far-flung legions for weapons, tools, metal parts for chariots, utensils, and, above all, money, the Romans transformed Tartessus into their most important source of metals.

Roman mining engineers introduced their shaft and gallery system of mining, and drove deep tunnels into the main ore body inside the coloured hills. An industrial operation of unprecedented scale was brought into being. According to the Roman historian, Pliny, there were between ten and twenty thousand slaves working in the mines of

One of the tunnels driven into the hillsides by the Romans during their 500 years of mining at Rio Tinto. *Few such examples of Roman techniques now remain, as modern operations steadily swallow up the earlier workings.*

Spain, perhaps half of them at Tartessus. Silver was still important, but by about the second century AD huge amounts of copper and growing tonnages of iron were also being mined.

An extensive programme of research at Rio Tinto over recent years by teams from the Institute of Archaeo-Metallurgical Studies in London has provided us with a detailed picture of Roman mining technology in action. The engineers had to solve a number of basic mining problems. One was the support of the roofs of the galleries. This was done much as it is today, with wooden pitprops. Many of these have survived, because they were more or less pickled in water containing copper salts.

The next problem was underground lighting. The Romans took care of this with thousands of terracotta lamps, burning good Spanish olive oil, placed at intervals of a few metres along the tunnels. Then there was the question of ventilation. The Romans not only dug air shafts, but lit fires at the bottom of them. The fires created strong updraughts, which sucked fresh air along connecting galleries into the deepest recesses of the workings. The ore itself was taken to the surface in small grass baskets, very similar to those still used in Spain today to carry farm produce.

The most serious problem, as with all mines throughout history, was drainage. Water is the ancient and universal enemy of the miner. By their nature, mines tend to fill with water by seepage from underground aquifers. Rio Tinto was no exception, despite the arid nature of the surrounding countryside. In the upper workings the Romans cut drainage adits or tunnels, which led the water out on to the side of the hills. But as their shafts went deeper they could no longer run the water out this way. However, in their efforts to get at the lower parts of the ore body they devised an extraordinary system of waterwheels.

These were made of wood, and were nearly five metres in diameter. Built into the rim of the wheels were wooden buckets for lifting water. The Romans installed pairs of wheels in the deeper tunnels, their lower rims in channels where the water collected. They were turned by slaves, either by walking on top of the wheels in treadmill fashion, or sometimes by pulling them round by hand. The scoops lifted water from the trough in the gallery floor to another trough in a higher tunnel. From here it was picked up by another pair of wheels, and again raised to a higher level, and so on until the water could be made to flow out on to the hillside along a drainage tunnel. Spanish miners pumping out an old part of the mine at Rio Tinto last century found the remnants of a nest of sixteen such wheels, arranged in pairs at eight different levels, providing a total lift of about thirty metres.

Some impression of this Roman technology can be obtained from the portion of a water-lifting wheel from Rio Tinto which is preserved in the British Museum, or from the almost complete specimen in the Huelva Museum in Spain. An even better demonstration is to be seen in the ancient Syrian city of Hama. Lining the river which runs through the centre of the city are the last remaining working examples of Roman water-lifting wheels, called *norias*. They are almost identical to those used at Rio Tinto, except that they are much larger – some

Water-lifting wheels, called 'norias', in the Syrian city of Hama, are direct descendants of the Roman wheels used underground in the Rio Tinto mine in Spain. There, turned by slaves, they were used to raise water from the deeper levels, as the Romans struggled to keep the mines drained. Here, turned by the flow of the river, they lifted water into aqueducts to supply the city. *The city of Hama, 200 km north of Damascus, still has nearly a score of norias working, although today they no longer supply water to the town. The largest is 20 metres in diameter.*

measure twenty metres in diameter – and are turned by the flow of the river itself. They lift water into aqueducts, which formerly supplied the city. The *norias* in Hama do not of course date from Roman times, although parts of them, including the great bronze bearings and hubs, may be four or five hundred years old. But they are absolutely unchanged in design, and as they turn majestically and unceasingly, with their characteristic groaning and splashing, they stand in direct line of descent from the slave-turned wheels in the dark, wet tunnels of Rio Tinto.

The peak of Roman trade, commerce, and the production of its huge mining operations at Tartessus was reached in the first two decades of the second century AD, during the reign of Trajan (who, coincidentally, was born in southern Spain). As the second and third centuries passed, however, the combination of inflation and declining economic fortunes led to progressive devaluation of the Roman currency. For a time this brought an increased demand for copper from Tartessus for coinage, which had previously been predominantly of silver. But when the Empire began to fragment in the fifth century, and the Visigoths sacked Rome in AD 410, the Romans withdrew from the Iberian peninsula.

Before considering the repercussions of the fall of Rome and its implications for our story, there is one strange episode in the chron-

A carved stone column in Rome is devoted to the victories of the Emperor Marcus Aurelius against invaders at the Danube. In its detailed depiction of the bronze and iron weapons borne by the legions, it is a graphic reminder of the 'Pax Romana' imposed by the world's first technological empire. *Like other detailed records of Roman activities, such as Trajan's column, this monument, after surviving for nearly 2000 years, is being rapidly destroyed by atmospheric pollution.*

ology of Rio Tinto to be told, and it is in some ways as extraordinary as any in its long existence. It concerns the way in which this great mine, perhaps the most renowned in the whole civilised world, was lost for a thousand years. It is related in David Avery's account of Rio Tinto, *Not on Queen Victoria's Birthday.*

Following the sack of Rome, the Visigoths went on to overrun Gaul and finally Spain. They are thought to have occupied the south about AD 475, but they could have had little interest in the extensive mine workings they found at Rio Tinto. They used very little coinage, and had no need of much copper. In any case, once the Romans stopped their drainage control measures the deeper shafts and drives would soon have flooded, submerging the waterwheels and drainage tunnels. And so the tenacious shrubs and grass began to grow back across the spoil tips and access roads. Rio Tinto slowly faded from human view.

In the eighth century AD the Moors swept into Spain from North Africa, and for six hundred years ruled it as an independent caliphate. The Moorish occupation lasted as long as the Roman, and was in many ways more creative, but there is no archaeological evidence that Rio Tinto was ever mined. The Moors were driven out of Spain in 1492, and the country returned to Catholic rule. That same year Columbus sailed from Huelva, at the mouth of the Rio Tinto, on his first momentous voyage in search of the riches of the East. But back in the coloured hills where the river rose, the forgotten workings and wealth of Tartessus lay undisturbed and unsuspected.

By the middle of the sixteenth century Spain had won a great empire in the Americas, and had absorbed, and spent, vast riches of gold and silver. Philip II found himself constantly short of money to finance the wars and intrigues that he and other Habsburg rulers were waging all over Europe. In 1556 he ordered a nationwide search for precious minerals and, in particular, for old mines which might be worth re-opening.

Leading one of the many search parties that went out to roam the remote hills and valleys of Spain was one Francisco de Mendoza, a member of the king's Council of Finance. Mendoza had been ordered to investigate some ancient Roman mines found in the province of Seville, and to look for others. In July he and a small band of men were exploring the mountainous country inland from Huelva. They were following up local village rumours of disused workings in remote parts when they came over a ridge 'and saw ahead of them a network of valleys in which great mounds of slag stood in dark contrast against the green shrub-covered hillsides. As they rode on they recognised with mounting excitement the signs of ancient Roman occupation: carved columns, the dressed stone of tumbled walls and immense drainage adits emerging from the hill-face.'

Mendoza and his party explored some of the galleries, and in them found mining hammers and picks, Roman lamps, even skeletons. When samples from the tunnel walls showed traces of gold, silver and copper they knew they had found what legend had promised but reason had despaired of – the lost mines of Tartessus.

The mines were reopened, and over the next three hundred years went on disgorging their wealth. But in 1873 the Spanish government, in need of cash, sold them to a British consortium. In its eighty-year tenure the Rio Tinto Company raised production to its highest level since the Romans left. (The British staff can claim one other not inconsiderable achievement: they introduced soccer to Spain.)

In 1954 the controlling interest in Rio Tinto returned to Spain. Today the still-expanding workings are steadily swallowing up all traces of Rio Tinto's many previous operators. But there is one small area on a breezy hilltop, surrounded by mine buildings and parking areas for huge earth-movers, where the past seems very close. It is the Roman cemetery. Shaded by dark, sighing pines, the graves of scores of miners lie undisturbed, beneath massive granite cover-stones whose inscriptions have long since weathered into anonymity. To stand among them is to reflect again on that endlessly fascinating subject of speculation: the reasons for the still-echoing fall of the Roman empire.

In the aggregate of factors which brought it about, the failure of Roman technology where mining is concerned is often overlooked. The Romans were great exploiters, but not great prospectors. For the most part, mining in the Roman empire was a depredation, with little thought of conservation or husbandry of resources. Conquest had no answer when the easily winnable ores near the surface had gone. The Romans could not cope with the invasion of water into the deeper mines. Yet they held the keys to essential knowledge which might have assured continuity. They knew about hydraulics, about gears, and understood the properties of steam. But they failed to create the new technology, the power pump, which would have allowed them to go deeper into the earth, where they knew, or must have sensed, that richer belts of minerals still lay. Rome retreated from the undergound frontier as well.

As the great architecture of the Roman empire began to crack and fragment, no part of its extensive geography was left unscathed by that slow erosion of will and capacity. The repercussions were felt far beyond the limits of the empire itself – as far as the crossroads of Asia, where trade between East and West had reached a new peak during the first centuries of the Christian era. If the emperors of Rome and China did not actually exchange ambassadors, they certainly knew of one another. Goods and ideas flowed freely between their two empires, carried by the ships of the desert, the slow but tireless camel caravans. From China to the West came a new and wondrous kind of cloth, which gave its name to the most historic of all trade routes, the Silk Road. From India came the diamonds, pearls, ivory and perfumes that were in such demand by luxury-loving Roman women. In exchange there moved eastwards a steady stream of Roman gold, much of which stayed in India, beginning the accumulation of wealth which would later help to found the princely states.

Between the great empires of East and West, many states, cities and peoples had fostered this trade. By maintaining the routes, guarding the passage of caravans, and guaranteeing safe conduct they had grown rich on the tolls they charged. One such way-station was Palmyra, in the Syrian desert. This settlement had grown up round a permanent oasis,

An indication of the luxury trade that passed through Palmyra, in the Syrian desert, is a delicate glass perfume bottle, possibly Phoenician, found in the ruins of this once prosperous city at the western terminus of the Silk Road from China. *Collection of the author.*

and was first known as Tadmor, city of dates. After the Romans occupied it in the first century BC the name was changed to Palmyra, city of palms.

Over the next four hundred years Palmyra developed into one of the most prosperous entrepôts of trade in the ancient world. It was the western terminus of the Silk Road, where goods from the East were transhipped for distribution all over the Mediterranean basin. The wealth that stayed in Palmyra enabled its citizens to indulge in much conspicuous consumption, especially in the form of public buildings. Surviving statues of the men of Palmyra, displayed in the National Museum in Damascus, show them staring coolly out at the world, confident of their economic security. Their women seem equally composed and are invariably shown festooned with jewellery.

But as both Roman and Chinese empires weakened, and barbarian pressures on the great overland routes increased, trade between East and West began to decline. The fortunes of Palmyra declined with it. In the third century AD the self-appointed 'Queen' Zenobia led an ill-judged rebellion against Rome. She was just half a century too early; the

Palmyra lies today in the desert sun, the skeleton of the once great Bronze and Iron Age civilisations of the Near East and the Mediterranean. With the collapse of the western Roman Empire, there was a pause of a thousand years before the centre of gravity moved, with the Renaissance in Europe, towards the north Atlantic. *Palmyra lies 210 km northeast of Damascus. Guarded by its isolation, it is almost unique among ancient cities in the Near East in having been subjected to no modern building or development.*

empire was not yet on its knees. Diocletian sent the legions to pillage and devastate the city. An earthquake completed the destruction, and Palmyra sank into an endless dream of the past in the Syrian desert.

Today the ruins of Palmyra rise in imposing array against the clear desert sky, uncluttered by modern buildings. In its great Temple of Ba'al, its Roman amphitheatre and arches, and its endless Greek colonnades leading nowhere, Palmyra provides the most vivid surviving glimpse of the ancient civilisations of the Mediterranean. Here are their skin and bones, bleaching in the sun. Among them it is possible to sense that implosion of mediocrity that overwhelmed the declining Roman empire, and brought to a halt the disciplined corporate endeavours which had advanced human society in so many significant ways.

The Dark Ages which followed did more than stifle learning and culture across the greater part of the Western world. It would be another thousand years before men would once again use metals on the scale that the Romans used them, either for the comforts of life or for the arts of war. It would need that millennium before the centre of gravity of the civilised world, which had moved with the Iron Age from the Near East to the Mediterranean, would move again, across the Alps to the lands bordering the Atlantic and the North Sea.

There, an ecumenical culture would spring to life, reviving the lapsed ingenuity of both East and West, and setting the peoples of Western Europe on the rise to their long ascendancy, and colonisation of virtually the entire globe.

CHAPTER 5
Shining conquests

The Bosporus at Istanbul is one of the most celebrated of the many shores of the ancient world, and perhaps its most vivid frontier – the one that separates Europe from Asia. For more than a thousand years, ships of all nations have battled the fierce currents of this narrow strait, which joins the Black Sea to the Sea of Marmara and the Aegean, while smaller craft plied across the Golden Horn, the narrow arm of the Bosporus thrusting into the city between the misty domes and minarets of the mosques and the huddled passageways of the bazaars.

The Bosporus is part of the geography of one of the most brilliant legends of history: the quest for the Golden Fleece. According to that Homeric tale, which has passed literally into every language, it was through the Bosporus that Jason sailed with his gallant band of Argonauts to what were then the secret ends of the earth, to retrieve the Golden Fleece. The legend itself has become a measure of heroic achievement, an allegory for the conquest of the impossible. But, like many legends, it may have been based on more prosaic human endeavours – in this case an expedition carried out by a group of determined men to seize a shining prize.

In the first millennium BC, around the Aegean, the most plentiful source of gold were the fine grains of gold dust mixed with river sands. The most effective method of retrieving alluvial gold was by washing the sands over sheepskins. The gold dust clung to the greasy fibres of the wool. When it could hold no more the golden fleece was dried in the sun and then incinerated in a fire. The grains of gold melted together into blobs, which could be recovered from the ashes. The richest sources of alluvial gold were the rivers of Anatolia which flowed into the Black Sea, and it was through the Bosporus to those distant regions that the Greek adventurers sailed in search of fortune as much as glory.[33]

The heroic legend of Jason and the Argonauts may therefore have been one of the first and most poetic records of an historic human

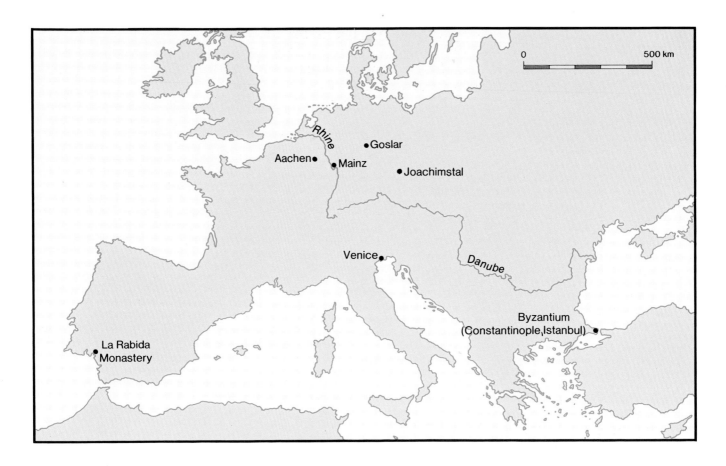

obsession. It is a compulsion that has driven man to the far corners of the earth, in search of that metal which above all has inspired his veneration and aroused his lust. That hunger for gold was an integral force in the great surge of expansion which followed the revival of Western Europe from the paralysis of the Dark Ages. It was gold as much as curiosity or adventure that drove the explorers and then the conquistadors across unknown oceans to the New World. And it was gold which in the nineteenth century set off the most feverish mass migrations in history, and which in the end helped to establish new nations in the southern hemisphere.

In the first few centuries of the Christian era, however, such expansionary prospects could hardly have borne contemplation. The fall of the Roman Empire had brought to a halt the technological and social advances which had raised the Mediterranean powers. There followed a stagnation of thought and enterprise in Western Europe, and a disruption of trade and contacts between East and West, and between Africa and Europe. Each region was thrown back on its own resources, to survive as best it could.

But it should not be imagined that this was a period of universal disorder. The fall of the Roman Empire was in fact no more and no less than the collapse of the Western Empire; the Eastern Empire not only survived but flourished. When Constantine established his capital in the Greek city of Byzantium on the Bosporus and renamed it Constantinople, the Byzantine empire entered a period of stability unmatched in

96

the Western world, and gained a pre-eminence that it was to keep for seven hundred years. Meanwhile, aroused by the exhortations of Mohammed, the Arabs burst out of the desert to impose the will of Allah in a great crescent around southern Europe, from the Black Sea to the Pyrenees. From the Germanic tribes which had overrun the Western Empire emerged the peoples who would lay the foundations of European civilisation: the Franks, the Anglo-Saxons, the Vikings. In the East, the T'ang emperors presided over an autumnal flowering of Chinese culture.

However, these bursts of energy were more or less isolated. Although there were vast and restless movements of people, there was no cumulative advance towards a new level of civilisation. Among the many reasons for this standstill, which lasted almost to the end of the first millennium AD, was one which is rarely considered in studies of the Dark Ages: the interruption to the supply of metals which had been caused by the fall of the Roman Empire.

In Europe, once the Roman underground techniques and organisational skills were lost, the mining of metals ceased almost entirely, except for the small quantities of easily accessible iron needed for weapons and tools. The once great mines of Rio Tinto and Cyprus were overgrown and forgotten. When the accumulated stocks of Roman coins finally became depleted, there were no precious metals to maintain a basis for currency and external trade. Coinage virtually disappeared. There was little buying or selling, even internally. Knights could not be paid, except in land. Farm workers were encouraged to farm some of their master's estate instead of receiving wages. It was the beginning of the feudal system.

Before this stagnant, closed economy could hope for any revival, or make any progress towards a higher standard of living, new supplies of metals, both precious and practical, would have to be found. Lost mining and metallurgical technologies would need to be revived. Only then would it be possible for Europe to acquire the financial and economic resources to generate growth and prosperity.

If the crucial revival of spirit, will and material enterprise can be said to have begun in any one place and through any one human inspiration, then it began in Aachen in Germany, under Charlemagne, in the ninth century AD. Charlemagne was the greatest of the Frankish kings. Having conquered almost the whole of Europe, from the Baltic to the Mediterranean, he converted it to his own brand of Christianity. Charlemagne was crowned emperor of the Western world by Pope Leo III in AD 800, and devoted his immense energies to creating a mixed culture to which all the disparate factions within his empire could contribute: Franks, Romans, Germans, Celts, Scandinavians and Slavs.

Charlemagne instituted drastic reforms of administration, finance and education which laid the foundations of European life as we know it. But he also awakened a renewed interest in metals and metal working. Goldsmiths and other craftsmen developed great skills. Besides the precious metals, others of more utilitarian character were sought out. Lead was mined and smelted in quantities unknown since Roman times, when it had been widely used for water mains, cisterns and pipes. Now it was in growing demand for new architectural features: the lead roofs and lead-lined stained glass windows of the great cathedrals which began to rise across Europe.

One of the most important factors in this revival of mining and metallurgy had been Charlemagne's decision to establish his capital at Aachen (Aix-la-Chapelle), midway between his Frankish homeland and the conquered territories of northern Germany. As a result, economic activity expanded east of the Rhine into Saxony, and beyond into Bohemia, Slovakia and Moravia, what is now Czechoslovakia. And this whole region, until then a sparsely-populated mountain wilderness, the haunt of wolf and bear, was found to be one of the richest mineral provinces in the world. From here, in the tenth century, would come the silver needed for the revival of commerce, and the copper, lead and tin for the growing industries that would depend upon them. It was the beginning of a new era in human history.

Charlemagne was quick to exploit the discoveries of minerals, using Saxon slaves to work the mines. Silver was recovered in such quantities that the Emperor was able to use it to renew the currency, and for centuries silver was the principal coinage of Europe. (One of Charlemagne's innovations was the pound, divided into two hundred and forty pence.)

By this time many people had learned to identify the outcrops of mineral ores that pointed the way to the riches beneath the earth. Prospecting had become a business. But then, as now, some of the greatest discoveries were made by accident. According to one story, still recounted in Germany, a noble out hunting in the Harz Mountains of Saxony in the year AD 938 tied his horse, Ramelius, to a tree while he stalked some game across a forested hillside. As horses will, Ramelius frequently pawed the earth while waiting for his rider. When the hunter returned he saw, in the hollow in the pine needles scooped out by the horse's hoof, a gleaming patch on the underlying rock. Miners were called and identified it as an outcrop of silver-lead ore.

The mine began as a small pit, but as the miners followed the vein down into the hillside, between walls of slate, it grew steadily thicker, until it joined a huge lode of silver, lead and copper, up to thirty metres thick, occupying the heart of the mountain. That mountain, and the mine, became known as the Rammelsberg, after the hunter's horse, Ramelius. It is still producing mineral wealth — now chiefly copper — after celebrating its 1040th anniversary.

In modern mining it is unusual for the methods and machinery of one period to survive the onslaughts of the next, but the Rammelsberg mine is an exception. Within the mountain many of the earlier workings still survive, and they provide a remarkable insight into the skill and determination of those Saxon miners.

They had no power tools or explosives, and the ore-laden rock of Rammelsberg was extremely hard, with few fractures or weaknesses. One method of mining, carried over from Neolithic times, was fire setting — that is, heating the rock face and then cooling it suddenly with water, to make it crack. It was common practice then for the miners to stay underground from Monday to Saturday, spending only

Sunday at home. So they would light big fires at the tunnel face on Saturday. By Monday, when they returned, the smoke had died down, and they could cool and fracture the ore, and break it up with crowbars. They hauled the ore to the surface in baskets, first using hand windlasses, and later horse-power.

But the deeper the miners dug into the mountain the bigger the challenge they faced from water. The history of Rammelsberg mining for a thousand years is a chronicle of ingenious endeavours to keep the workings drained. After the initial open-cut mining in the tenth century, the miners had begun to sink shafts and drive tunnels to reach the ore body inside the mountain. By the beginning of the twelfth century they were halted by the seepage of water into the shafts. And so began, in 1150, the driving of the first major drainage adit. This was a tunnel designed to run with a slight down-slope from the bottom of the main shaft to an exit on the side of the Rammelsberg. As fast as water collected in the mine it would flow out.

At that time the only mining tools for tunnelling through hard rock were hammers and chisels, and the rate of advance was never more than twenty centimetres a day, less than the length of a miner's forearm. It took thirty years to complete the nine hundred metre tunnel. Once the workings went below the level of the drainage adit, however, the inflowing water had to be lifted to the adit level before it would run away. At first, human labour was used to carry water up ladders in leather buckets. But such methods could not keep pace with the rising waters, and in the middle of the thirteenth century the water level reached the drainage adit, and mining had to be abandoned.

Nearly a century elapsed, and then in 1370 work was resumed, with the introduction of the first mechanical water-lifting system. This was a large treadmill, worked by men, connected by an axle to a large wheel located over the main shaft. This wheel operated an endless rope carrying inflated leather bags. The rope and its bags were enclosed inside a series of hollow tree trunks, which formed a vertical wooden pipe. The bags moving up through this pipe lifted water to the level of the drainage adit.

But the richness of the ore body drew the miners still deeper, and a new approach to the drainage problem was needed. In 1486 they began to tunnel a second drainage adit, some forty-five metres below the first. This was much longer – two thousand six hundred metres in all – and it took ninety-nine years to complete. Meanwhile, a dam was built in a high valley near the upper entrance of the mine. Water from the dam was led through tunnels to a series of waterwheels, assembled deep inside the mountain. These wheels operated a linkage of wooden beams and pumps, to raise water from the deepest levels to the new drainage adit. This also carried away the water which operated the wheels themselves.

With this solution to the flooding problem the Rammelsberg miners, so it is said, 'swam in silver' as they delved into the lowest levels of the ore body, three hundred metres below the mine entrance. Their remarkable waterwheel system was in use until 1910, and the last wheels to be installed, one hundred and fifty years ago, are still in

The Renaissance in Europe was made possible to a large degree by the wealth which flowed from the silver mines in Central Europe, such as the Rammelsberg mine in the Harz Mountains of Germany, which has been in production since AD 938. *Michael Charlton with one of the underground waterwheels used from medieval times until the beginning of the present century to drain water from the Rammelsberg mine.*

place. The largest is nine metres in diameter, and is still connected to its elaborate system of wooden cranks and rods, which disappear down the old shaft.

Within a few decades of its discovery the Rammelsberg mine became the most important source of silver, copper and lead in central Europe, and a key training school for miners, prospectors, and smeltermen. Its discovery attracted a flood of miners and fortune-seekers, closely followed by farmers, craftsmen, and merchants. The Harz Mountains became the centre of an ever-expanding region of growth and prosperity, as rich new deposits of silver and other metals were made in Saxony, Bohemia and Moravia.

The great landowners of the time were always pressed for revenues, and willingly threw open their estates to prospectors, especially to the Saxons, who were becoming talented at looking for minerals. Many would-be miners were released from feudal bondage in order to help develop the mineral wealth of the region. A prospector who found a deposit could mark his claim with piles of stones at each corner and register it with the landowner. For a tribute or royalty he could then mine the claim and sell any minerals he found. The miner had ceased to be a serf; he was at last a free man, his own master. So were laid the foundations of the mining law that was carried by the Saxons to Britain, and eventually by the Cornish miners to America, Australia and South Africa during the gold and diamond rushes of the nineteenth century. It is still the basis of mining law in most parts of the world.

The period from about AD 900 to 1200 in western Europe saw the slow lifting of the despair that had shadowed the Dark Ages. Pessimism gave way to optimism. The acceptance of poverty was replaced by the

The town of Goslar grew up alongside the Rammelsberg mine, and is perhaps the oldest mining town in the world. It has been preserved almost intact from the encroachments of the modern world, and in its quiet streets the Middle Ages never seem far away. *The Rammelsberg mine is finally nearing the end of its life, but Goslar appears to have an enduring future as a tourist centre.*

hope of improvement. This change of attitude was encouraged by the gradual cessation of raids by the Vikings from the north, the Saracens from the east, and the Hungarians from the south. The new social order was reinforced by the growing influx of silver into the economy from Rammelsberg and other mines.

The impact of that mineral wealth on the development of Europe in general, and of Germany in particular, can be sensed even today in the picturesque town of Goslar, which was founded at the foot of the Rammelsberg in AD 922. With its winding, cobbled streets and elaborately decorated buildings, Goslar retains a pervasive medieval presence. Here it is not hard to visualise the new form of society which emerged

101

from the long night of the Dark Ages. It was perhaps the first kind of society that we could recognise and perhaps identify with: a community of solid citizens, living in a well-built town, going about their affairs and minding their own business. In Goslar, the fruits of shrewd enterprise and civic virtue have been invested since those medieval times in houses and public buildings that have lasted the best part of a thousand years already, and look like lasting as long again.

The prosperity of Goslar has always been associated with the Rammelsberg mine, and there is a reminder of this on one of the buildings overlooking the town square. On the striking of an elaborate mechanical clock, of the kind so popular in Germany, a procession of miners with their tools emerges from one door, passes slowly in front of the clock, and disappears into the façade of the building again.

A more evocative symbol of the wealth that flowed from Rammelsberg into the economies of western Europe is to be seen in another building that overlooks the square. This is the fifteenth century *Rathaus* or town hall, with its extraordinary inner Hall of Homage. From 1500 onwards this room was the meeting place of the city council, but during some period of civil disorder its doorway was bricked up and forgotten. The room was rediscovered during renovations in the nineteenth century, when its elaborately painted ceilings and walls and their superbly gilded ornamentation were revealed in perfect condition. The paintings, depicting Biblical scenes in glowing tempura pigments on wood panels, embellished with gold leaf, are among the finest surviving examples of medieval decoration. They had been donated to the town by the Rammelsberg mine owners, and are an indication of the wealth that stayed behind in Goslar.

Most of the Rammelsberg's silver was sent out of Saxony, and became a hidden but indispensable factor in that most spectacular expression of European resurgence, the Renaissance. It enabled the Emperor Frederick II, among others, to become a noted patron of literature and science. By this time, other discoveries were adding to the growing prosperity. In 1516 one of the greatest silver strikes in history was made near the town of Joachimstal (now in Czechoslovakia). At peak production the mines disgorged three million ounces of silver a year. Much of it was minted in the town, into a coin known as the *joachimstaler*, or *thaler* for short. This is the word from which dollar is derived.

The mine owners became the financiers and creditors of the courts and princes of Europe. Henry VIII of England was one monarch who obtained loans from this source. The wealth from the mines was funnelled through the emerging centres of finance, and Venice in particular. It was the flood of silver from central Europe which was largely responsible for the transformation of this poverty-stricken, mosquito-ridden ninth century fishing village on the Adriatic into the richest port in southern Europe.

In the thriving world of Italian commerce the most important people were the money changers. Because they kept their cash in piles beside them on benches or *banques*, they became known as bankers. By the middle of the thirteenth century the bankers in Florence and Genoa were trading currencies, issuing bills of exchange, and speculating in exchange rates.

The prosperity which began to infuse many countries in western Europe not only created new classes of artisans, craftsmen, successful merchants and comfortable burghers, but gave opportunities for fertile minds to revive forgotten technologies from both East and West, and to combine them in technical advances which would transform the world. One such advance was printing.

In the long march of the common man, few weapons have proved as powerful as the book. Nothing in history has been more reviving to the spirits of the oppressed than the might of the printed word. The capacity to store knowledge and experience and to disseminate it widely and generously led to that explosive spread of literacy and learning in the fifteenth century which made printing one of the greatest concepts of civilised man. It was an overwhelmingly important factor in the evolution of present society.

The elements which produced that advance were known and available in many parts of the world long before it took place. They were tributaries of existing knowledge, but they came together to form their irresistible current for the first time in Mainz, in Germany, in the mind of one man. His name was Johannes Gutenberg, and he had so acute a perception of an age-old problem that his solution would require virtually no change for another five hundred years. It was a problem that had to await the attention of someone who reflected the growing interest in Europe in metals and metallurgy.

Before Gutenberg there were, of course, many ways of recording and storing knowledge. For thousands of years, people had incised clay tablets, carved ivory and wood, and drawn and written on papyrus and parchment. In Europe, by the Middle Ages, there were books of exquisite detail and beauty, but they were not for everyman. They were laboriously inscribed and painted by hand on sheets of split and dried animal skins – rough parchment from sheep and goats, and the finer vellum from calves. Because of the labour cost of producing books in this way they were necessarily available to very few. Before books could become freely and widely circulated, it would be necessary to find a cheaper material on which to print them, and a much faster way of reproducing pages of text or illustration.

Long before Gutenberg devised his metal type, documents were being printed in China from movable wooden blocks. *National Historical Museum, Peking.*

103

Both solutions came, like so many other technological advances, from the East. The Chinese, in the first few centuries AD, had discovered how to produce a much better writing surface than parchment or vellum. They used various plant materials, chopped up so that their fibres were reduced to tiny shreds. These were mixed with water in an open tub to form a milky slurry. When a fine screen was submerged briefly in the water the fibres settled in a thin layer on the surface of the screen, their short lengths randomly aligned so that they became naturally interwoven. When the screen was lifted out, the layer of fibre could be peeled off. When dried it formed a perfect surface on which to write, draw or print.

Knowledge of the art of paper making seems to have reached Europe in the eighth century AD, through the Arab world. The Chinese had apparently sent a team of paper makers to the city of Samarkand, in central Asia, just before that city was overrun during the great expansion of the Moslem empire. The Arabs made good use of their captured technology, and began making paper in Baghdad in the reign of the Caliph Haroun el Raschid. Arabian paper was soon circulating all through the Near East and Mediterranean countries. The first paper produced in Europe was made by the Arabs in Spain. By the thirteenth century paper mills were operating in Italy, using torn linen rags as their base material.

There was a great demand in Europe for the new writing material by the burgeoning financial industry, for the keeping of accounts and bills of sale. It was also used in universities for the copying of learned books and manuscripts. But although by this time paper had become much cheaper than animal skins there was one other factor which restrained the wider distribution of knowledge by the medium of the written word. That was the high cost of the scribes who did the copying by hand. The Great Plague had greatly reduced the populations of European cities, including the numbers of the relatively few who could read and write. Those who were left could charge extremely high prices for their work. What was needed was some repetitive and preferably mechanical method of reproducing text.

The inspiration came, again, from the East. The Chinese had long been printing on paper with characters carved from wood, and had also tried forming them from clay. However, such materials quickly wore out. The Koreans went further. In 1409 the first document known to have been printed from individual characters was produced in the state printing shop in Seoul. Studies of this and other Korean printing of the period suggest that the characters may have been made of metal, but none has ever been found. In any case, Eastern attempts at printing with individual characters did not develop – perhaps because many Eastern languages consist of ideograms, representing multiple words, rather than separate letters, which are more versatile.

Printing from wood blocks was well known in Europe by Gutenberg's time. This was the method used to produce playing cards. There were even metal stamps for making coins, and for inscribing characters on cast bronze bells. But it took Gutenberg's insight to

Opposite: One major contribution to the development of printing and the spread of knowledge through books was the invention of paper-making. Paper was first made in the Far East, by dipping a screen into a slurry of plant fibres in water; the screen picked up a thin mat of interlaced fibres, which dried to form a smooth sheet. *Japanese paper-maker at work near Tokyo.*

In China, blocks for printing were cut from hard, dense-grained pear wood, but even these wore out under continuous use. *The oldest surviving wood-block printery in Peking has been in production for more than 300 years.*

combine this technology into a system of printing on paper with movable metal letters.

Gutenberg was born around 1397, the son of a prominent official in Mainz. From his father, who had worked in the mint, Gutenberg learned much about the properties of various metals and their alloys, and for a time worked as goldsmith. He was a religious man, and spent many hours in the Carthusian monastery in Mainz, watching the monks inscribe their Bibles by hand with quill pens, a task which often took years. It may well have been Gutenberg's desire to help spread the word of God which finally turned his mind to the problem of printing.

Gutenberg must have realised that the solution lay in using separate, movable letters that could be arranged to form words. They would need to last longer than wood, and yet it would be much too expensive to have them laboriously cut from steel, as the stamps of the coin makers and bell casters were. And so Gutenberg had his great inspiration, which was to produce the letters by one of man's oldest techniques of metallurgy – casting.

For his casting metal Gutenberg devised a mixture of lead and tin which had a very low melting point, about 240°C. This alloy was not only quite easy to maintain in a molten state, but when it was poured into a mould it set instantly, losing sufficient heat to enable it to be handled almost immediately. To make the individual letter moulds Gutenberg first cut a steel stamp of each letter, and with it punched an impression into a small block of soft copper. This fitted into a small hand clamp, with a narrow space above the impression of the letter. When the space was filled with molten type metal it produced a cast

106

letter on top of a stem. Gutenberg could make as many copies of each letter as he desired, and all would stand the same height.

Gutenberg did not just devise a method of making metal type, but worked out a complete system of printing. He made a type case with many pigeon-holes, and kept each letter in its own hole. Picking letters from the type case, he assembled words along a grooved stick to make a complete line. A notch cast into the stem of each letter told him by touch which way round the letters should go in the stick, without his even having to look at them. The lines of type making up a page were locked into a frame, or *forme*, and inked ready for printing.

If all this sounds familiar, it is because Gutenberg's system is still in use in many parts of the world, basically unchanged since the fifteenth century. Although type-setting and printing from so-called 'hot metal' is being rapidly replaced in many Western countries by the cold green glare of computer-generated photoprinting, it will be a long time before metal type finally disappears.

To obtain a firm, even impression, Gutenberg adapted a device that had been in use in Europe since Roman times: the wine press. He placed a sheet of paper over his bed of type, covered it with a pad to hold the paper flat, and slid the *forme* under the press. When he tugged a handle projecting from a vertical, screw-threaded wooden shaft, the screw exerted downward pressure on the sheet of paper, producing a perfect impression of the page of type. For the first time there existed a means of reproducing and disseminating knowledge on a universal scale. At a stroke, all human wisdom, imagination, folly and achievement could be drawn upon.

Gutenberg's first creation was, however, the Bible. For his type face he copied the 'textura' script used by the monks of Mainz, despite the problems he had in reproducing the double letters, inter-connections between letters, and abbreviations. Only in this way, he believed, could he print books which would be virtually indistinguishable from those written by hand. And such was his objective.

It took Gutenberg three years, from 1452 to 1455, to print about two hundred complete copies of the Bible for sale. It was slow and expensive work. He printed thirty copies on vellum, which required the hides of almost thirty thousand calves. The remaining copies were printed on handmade paper, prepared from shredded cloth. Records maintained in the Gutenberg Museum in Mainz show that forty-eight copies survive today, in various parts of the world. Some are incomplete, but the register carefully lists the present state of every surviving copy. Even single pages are recorded.

Three superb copies of the Gutenberg Bible are on display in the museum in Mainz. These underline one astonishing feature of the first major work to be printed from movable metal type. It is that each page of each copy is as perfect as any other. The printed words, interspersed with coloured designs, are all immaculately aligned and reproduced, completely free from smudges, blots, or alterations. Somehow, this extraordinary work was brought into existence fully conceived. All the components of Gutenberg's system appear to have worked faultlessly from the very beginning, with no evidence of trial and error, or failures of technique. It is this initial perfection which still raises some doubts about Gutenberg's originality, and suggests to some authorities that he must have drawn upon other people's earlier and perhaps less successful experiments. There is, however, very little evidence which detracts in any way from Gutenberg's great achievement.[34]

Unfortunately, the Gutenberg Bible was the first and last masterpiece of a man who soon lost everything in debts and law suits. He lived on patronage until his death in Mainz in 1468, destitute, blind, and forgotten. But history attributes to Johannes Gutenberg a prophetic insight. 'With my twenty-six soldiers of lead,' he is reputed to have said, 'I shall conquer the world.'

How right he was. Gutenberg's lead soldiers marched down all the roads of Europe, and almost all the roads on earth. The enormous stream of books which poured off the printing presses in the wake of his Bible promoted the spread of knowledge into all levels of society. By 1500, it has been calculated, some thirty-five thousand separate editions of books, totalling perhaps twenty million copies, had been distributed. It was the greatest cultural revolution since the invention of writing. The past was brought into the present. That which was remote was brought near. The whole of mankind was made vocal. Long forgotten heroes now spoke to people everywhere. For the first time, there was some guarantee that the great concepts of the human intellect would never again be lost, as so many of the classical texts were lost after the fall of Rome and the burning of the library of Alexandria by the Arabs.

One of the most immediate benefits of Gutenberg's invention was a tremendous exchange of technology in almost every field of human activity. In 1530 the town physician of Joachimstal, the famous silver-mining centre in Bohemia, gave up his job and devoted himself for the next seventeen years to the collection of all available knowledge about mining. In 1556 he published *De Re Metallica* (*On the Subject of Metals*), under his pen-name, Georgius Agricola. This comprehensive work, the first of its kind on the finding, recovering and processing of metals, not only became the miner's bible for the next two hundred years, but underlined the crucial role being played in the renaissance of Western life, culture and commerce by the continuing infusion of wealth from the mines of central Europe.[35]

While silver breathed new life into trade, it also stimulated the new preoccupation of the European powers, the conquest of distant regions of the world. Among the factors which fuelled this expansion were critical improvements in the design of ships and sails, which made long sea voyages practicable. The technical advances included the boxed compass and the stern-post rudder, which enabled ships to hold a course into the wind. There were now the financial resources to equip and sustain long expeditions. And there was one overriding impulse: a hunger for territories and subjects, for slaves and spices, and above all for treasure. The extension of European influence around the world in the fifteenth and sixteenth centuries was inseparable from the thrust after mineral riches which alloyed wealth with power.

The most celebrated search launched by the Western world for the

GEORGII AGRICOLAE

DE RE METALLICA LIBRI XII ▸ QVI▸

bus Officia, Inftrumenta, Machinæ, ac omnia deniç ad Metalli-
ram fpectantia, non modo luculentiffimè defcribuntur, fed & per
effigies, fuis locis infertas, adiunctis Latinis, Germanicifç appel-
lationibus ita ob oculos ponuntur, ut clarius tradi non poffint.

E I V S D E M

DE ANIMANTIBVS SVBTERRANEIS Liber, ab Autore re-
cognitus: cum Indicibus diuerfis, quicquid in opere tractatum eft,
pulchrè demonftrantibus.

BASILEAE M ▸ D ▸ LVI ▸

Cum Priuilegio Imperatoris in annos v.
& Galliarum Regis ad Sexennium.

The title page of *De Re Metallica*, the first major work on mining and metals, published in 1556 by Georgio Agricola. It was translated from the original Latin into English in 1912 by Herbert Hoover (later President of the United States) and his wife, Lou Henry. *From a facsimile page in the unabridged edition produced in 1950 by Dover Press, New York.*

Above: An illustration from *De Re Metallica*, showing early attempts to solve the problem of water in the mines. *From the Dover Press edition, New York, 1950. Above right*: *De Re Metallica*, with its illustrations of mining techniques, became the miner's bible throughout Europe for the next two centuries. *From the Dover Press edition, New York, 1950.*

rumoured mountains of gold and silver that lay beyond the setting sun began at the little monastery of La Rabida, on the flat estuary of the Rio Tinto, near Cadiz in Spain. It was here that the expedition was planned by a man who claimed to be 'divinely selected': Cristobal Colon, better known as Christopher Columbus.

All explorers need visions, it has been said, as lesser men need food, and Columbus certainly had his. It was based on Marco Polo's description of China as a country 'richer than any other yet discovered', possessing gold and silver, precious stones, and all kinds of spices, 'things which do not reach our country at present'. Columbus was seized by the idea, incessantly talked about in the seaports of Europe, that a ship sailing westwards would reach, eventually, India and China. But he got no practical support for his dream of making just such a voyage until he chanced to call with his son, Diego, at the monastery of La Rabida in 1485.

The monastery still stands, virtually unaltered, and is today maintained by the monks as a shrine to those distant, momentous events. It is a low building of whitewashed stone, with a red tiled roof, surrounded by tall palms. It is now separated from the sluggish course of the Rio Tinto by a stretch of marsh, but in the fifteenth century the river bank was just below the wall. Inside the monastery there are courtyards shaded by ancient orange trees and bougainvillea. Many of the rooms have been preserved exactly as they were when Columbus and his son were welcomed by the friar, Father Antonio de Marchena. The scene of their meeting in the courtyard is shown in a picture in the monastery, painted some two hundred years after the event.

Father Marchena was an amateur astronomer, and had a wide interest in the world. This may explain the warmth of his welcome to Columbus, and the sympathy he showed to his visitor's obsessive idea, which emerged as they sat and talked in a little room which is now known as 'the Bethlehem of the Americas'. In the end, Father Marchena offered to help Columbus by recruiting crews from the local communities, renowned for their seafaring traditions. And so it was that the monastery of La Rabida and its monks provided Columbus's main support and encouragement during the seven years that it took for his dream to become a reality.

The altar in the chapel of the monastery of La Rabida, at the mouth of the Rio Tinto in Spain, where Columbus made his final prayer before setting out on his historic voyage of discovery to the New World. He asked God, he wrote later, to show him 'where gold is born'. *The monastery and the chapel, with its figure of the Virgin of Vilagros above the altar, is maintained by the monks as 'the Bethlehem of the Americas'.*

111

Finally, in July 1492, Columbus found himself at the monastery once again, making plans for his voyage with Father Marchena and the monks, and the captains of his three ships. Columbus and his men spent some weeks in residence, dining with monks, servicing their ships, endlessly studying maps and globes, and preparing themselves for the vast unknown. Columbus himself walked the corridors and courtyards, convinced, as he later wrote to his patrons, King Ferdinand and Queen Isabella, that 'neither reason nor mathematics nor maps' would be of any use to him in his enterprise.

It was in the little chapel of La Rabida, before the figure of the Virgin of Vilagros which still glitters in the candlelight, that Columbus made his final prayer. It was coupled, he wrote later, with an important request: 'I asked God to show me where gold is born.' And so on 3 August 1492, Columbus's men filled their water casks from the well which still stands near the river, and at eight in the morning the *Santa Maria*, the *Pinta* and the *Nina* cast off and sailed down the Rio Tinto and out into the Atlantic.

In the scale and dimension of its consequences, and because it changed so many things for so many people, Columbus's first voyage may be regarded as one of the most important journeys that man has made. There was an air of expectation over the enterprise. It was time for the Europeans to make that voyage. They had the ships, they had the men, they had the incentive, and above all they had the money. It was time to go.

What followed was, in the broadest sense, the dawn of global civilisation. Until the end of the fifteenth century, human culture and society had been essentially land-centred and land-linked. Contacts by sea had been peripheral and uncertain. Now, direct sea contact between far distant continents was about to be established. Vast regions and entire peoples who had formerly gone their own way in isolation would be brought on to the stage of world history. It was the beginning of the challenge to the old axis of civilisation between Europe and Asia.

The story of the Spanish conquest of the New World is an extraordinary and many-sided one, and it does not concern us here, except in one important respect: it was undoubtedly the lure of gold as much as anything else that led to the discovery of the Americas. Once established there, Spain mined with the sword rather than the pick, as Columbus was followed by the conquistadors.

The conquistadors believed themselves to be the chosen instruments of Almighty God, and their overwhelming statements in the New World still stand to affirm it – the mighty cathedrals which dominate towns and cities from Mexico to Peru. None speaks more powerfully for that turbulent period than the Cathedral of Lima, for it contains the mummified body of Francisco Pizarro, the conqueror of the Incas and perhaps the most obsessively determined of all the conquistadors. A huge bronze statue of Pizarro on his horse, sword upthrust, his face hidden beneath a menacing, visored helmet, stands near the cathedral, in the square of the city that he founded in 1535, baptising it with the ringing name of Cuidad de los Reyes – City of Kings.

In the main square of Lima, with its still dominant Spanish presence, it is not difficult to imagine the mixture of excitement, exaltation and fear that must have filled the minds of the conquistadors as they came marching into South America. The pyramids of human sacrifice which confronted them in Aztec Mexico had done much to enflame their conviction that they were men of mission, with a mandate to enforce righteousness, as well as pursue riches, by the sword. With that knowledge, and their own ruthless audacity, they were twice armed. The treasure which Montezuma offered Cortez in propitiation had served only to whet more predatory appetites. Before them lay a prospect of riches that outshone any in the legendary past.

As Pizarro and his little band – one hundred and ten on foot, and sixty-seven on horseback – marched up the high valleys of the Andes they ransacked the Inca empire in one of history's more single-minded conquests. They harvested gold. One massive prize came in the form of ransom for the captured Inca ruler, Atahualpa. The price offered by Atahualpa was a large roomful of gold objects, piled as high as a man could reach. From all over the empire, litters laden with gold articles of every kind were brought in to make good the ransom – goblets, jugs, vases, trays, utensils, plaques from buildings, ornaments, figures of animals, golden ears of corn sheathed in leaves of silver, with tassels of fine gold chains. When it was all handed over, Atahualpa was killed. He escaped being burned at the stake only by accepting baptism, and was then garrotted.

The final chapter was played out in Cuzco, the Inca capital in a high, open valley, which Pizarro and his men entered on 15 November 1533. Here the Spaniards, having razed the Inca civilisation, set about reducing their mountain of booty to manageable form. There were no easily transportable coins or bars. The Peruvian people had made no use of

The massive stone structures built by the Incas of Peru in the high valleys of the Andes are tangible reminders of the empire destroyed by the Spanish conquistadors, in one of history's most tragic treasure hunts. *This wall, made of huge granite blocks fitted together without mortar, stands above Cuzco, the former Inca capital, which was entered by Pizarro and his men in 1533.*

The spectacular ruins of Machu Picchu, the secret city of the Incas among the peaks, which the Spaniards never found. It was eventually abandoned, for reasons which have never been established, and was lost to the outside world until re-discovered in 1911. *Machu Picchu is still accessible only by a narrow-gauge railway from Cuzco, built originally by a mining company, which winds through the steep valleys.*

gold or silver for currency, or as a measure of value. For them, gold was a symbol of grandeur, directly related to their concept of divinity. It was essential in rituals, but equally useful for ornaments and objects of daily use. So the Spaniards began the task – which took them many months – of melting down into ingots the craftsmanship of centuries, the greatest accumulation of treasure ever seen.

In the end, the wealth of Peru became a source of discord, and a curse. In the years that followed the Spaniards fought among themselves with deadly ferocity, until Pizarro himself was assassinated by some of his embittered followers. But the influence of the conquistadors was to last long after their shining conquests were over. On the rubble of the Aztec pyramids and the Inca palaces Spain raised up the might and majesty of her Catholic empire, whose spirit persists so tenaciously throughout her former territories.[36]

After the fall of Cuzco, the remnants of the Inca forces retreated to their last strongholds in the mountains. Some, like Machu Picchu, were so inaccessible that the Spaniards never found them. This astonishing city of stone, built with consummate architectural skill on a steep-sided pinnacle nearly three thousand metres above sea level, was lost and forgotten, along with the Incas themselves. Machu Picchu lay undiscovered and unsuspected for nearly four hundred years. It was only in 1911 that the American explorer, Hiram Bingham, found its moss-encrusted ruins beneath a mantle of ferns and creepers.

Archaeological studies have peeled away some of the mysteries surrounding this human roost among the cloud-topped peaks. All the evidence found during excavations – ceramics, metal objects, weavings – suggest that Machu Picchu was built around 1420, during the reign of Pachacutec, the ninth emperor of the Inca dynasty. It was the climax of a vast construction programme, which included the re-building of Cuzco, demolished by an earthquake.

Using stone hammers and bronze crowbars – the Incas had no iron – the workers quarried stone and shaped it into blocks, which they fitted together so perfectly that mortar was rarely needed. Some of the massive stone lintels over the doorways weigh many tonnes. The cutting of the rock steps that link the various levels of the city was a huge undertaking in itself. There are more than one hundred staircases, containing three thousand steps. Within the perimeter, which measures about five kilometres, there are two hundred and fifty buildings, including palaces, temples, towers, tombs, houses, and granaries for storing the crops grown on the steep, terraced slopes. Water was channelled into the city from springs, and distributed by fountains. The total population was probably about one thousand.

What was the purpose of Machu Picchu? Most authorities seem to think that its primary function was religious, that it was built as a kind of convent for a group of women called 'Virgins of the Sun'. They were chosen from the Inca ruling class for their talents and physical perfection, and spent their lives in isolation, practising religious rites. This theory would explain why more than seventy-five per cent of the human remains found at Machu Picchu were female.

116

The end of Machu Picchu remains as clouded with mystery as its origins. Overlooked by the Spaniards, the city continued to exist at least until the beginning of the seventeenth century. But then, for reasons which have never been established, its life came to an end, and its population vanished. Machu Picchu sank into total obscurity for more than three centuries, until Bingham was led to it by some Indians who lived in a valley nearby.

Dr Hiram Bingham was an American teacher who had become involved in an obsessive search for the legendary Inca city of Vilcapampa. This was reputed to be the sacred capital of the Incas, a replica of Cuzco somewhere in the mountains, where the temples and palaces were faced with gold and silver. Although the Spaniards searched for it, as they destroyed the last vestiges of Inca power in 1571, they never found it. After the several expeditions and four years' work that it took to reveal the splendours of Machu Picchu, Hiram Bingham believed that he had found Vilcapampa. Whether or not this is true, the astonishing eyrie that he discovered remains the most evocative reminder of a civilisation which above all others in history had made gold an adoration.

Most of the gold objects seized by the Spaniards in South America were melted down, and of those taken back to Spain very few survived. Fortunately, over the centuries many exquisite examples of gold working have been found in graves and similar hiding places, and superb collections have been accumulated in the lands of their origin. Tangible evidence of the mystical regard that the South American Indians had for gold can be found in the two most important repositories of pre-Columbian culture – the Oro del Peru Museum in Lima, and the Museo del'Oro maintained by the Bank of the Republic in Bogota, Colombia. From these collections it has been possible to assemble a picture of a remarkable chapter in the history of metallurgy, which saw the noble metal used on a scale and with an artistic expression that has never been surpassed anywhere in the world.[37]

One intriguing finding is that, contrary to the direction of human occupation of the Americas, knowledge of gold working spread from south to north. The earliest use of beaten gold has been dated to about 2000 BC, in southern Peru. After that, for perhaps two thousand years, there were purely local developments in Peru, involving other metals, including silver, copper and platinum. Then, around the time of Christ, advanced gold technology, including lost wax casting, appeared in Colombia, on the northern coast of South America. Gold working reached Panama in the fifth century AD, and by about the eighth century it was flourishing in Mexico.

In Peru, before the Spanish conquest, copper was possibly the most common metal in use, but the next most abundant was gold, followed by silver. Meteoric iron was rare, and iron smelting was unknown. Nearly all the gold was alluvial, recovered from placer deposits in rivers. Before the grains of gold could be used they had to be melted together in clay crucibles. The Peruvians had no bellows, but used blow pipes to raise their furnace temperatures. The most widely used technique of gold working was the beating of gold and gold alloys into sheets, which were cut and shaped to make the desired objects. Casting was little used.

Indian metal-working technology reached its pinnacle in Colombia. In this part of South America gold was more plentiful than any other metal. Copper existed in limited distribution, and silver was generally found only as a natural ingredient of gold. Platinum was fairly easy to recover, from river sands, but because of its very high melting point – 1775°C – it was difficult to use. Tin was almost totally absent from this area, so the Colombians were never able to develop a bronze tradition. Instead, they adopted as their utilitarian metal the alloy of gold and copper, often with traces of silver, called 'tumbaga'.

This alloy of thirty per cent gold and seventy per cent copper had several advantages. It melted at 800°C, compared to 1064°C for pure gold and 1084°C for pure copper. It was easier to cast than either of its constituents, and reproduced fine decorative detail with more accuracy. Finally, tumbaga was harder than either gold or copper, and could be cold-hammered to produce tools almost as tough as bronze.

One of the great arts of the Colombian gold workers was in making objects of tumbaga look like pure gold. They did this by the sophisticated process of depletion gilding. Unlike the more familiar form of gilding used in the Old World, which consisted of adding or depositing a thin sheet of gold on the surface, depletion gilding created a pure gold skin from the gold already in the tumbaga. The gold workers did this by soaking the object in various substances obtained from plants and minerals, whose chemical constituents etched away the copper and silver molecules from the surface layer, but left the gold molecules untouched. After several such treatments there formed a thin coating of pure gold over the tumbaga base.

Another remarkable accomplishment of the Colombian metallurgists was their use of platinum, perhaps earlier than anywhere else in the world. Platinum occurs as tiny, grey grains of pure metal in the gold-bearing sands of the rivers draining into the Pacific. But because this element does not alloy in nature with gold, it came as a surprise to modern metallurgists to find that some cast gold items of Colombian jewellery contained platinum in proportions ranging from twenty-six to seventy-two per cent, in a perfectly homogeneous alloy. In theory, this was impossible, because there was no method by which the Colombians could have achieved the melting point of pure platinum, which is 1775°C.

Research has provided the only possible explanation. The workers heated grains of platinum with a small quantity of gold dust, until the gold melted and bound the platinum granules. In such circumstances, although the melting point of platinum was never approached, a little of each metal dissolved in the other. The mixture was allowed to cool, hammered to help blend the gold and platinum, and then re-melted, and beaten again. After prolonged treatment in this way, the result was an alloy of perfectly blended gold and platinum, which melted at a temperature below that of pure platinum, and which could be cast or worked like any other alloy of gold.

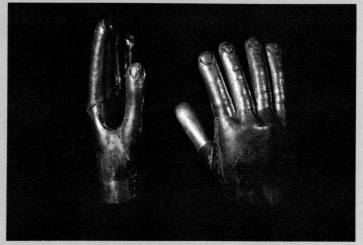

The greater part of the gold treasure seized by the conquistadors in South America was melted down and shipped back to Spain, but sufficient examples survived to convey some idea of the skills and artistic imagination of the Indian goldsmiths. *These specimens are in the Oro del Peru Museum in Lima.*

The most outstanding expression of Colombian craftsmanship in metal was their casting of gold by the lost wax process. They raised this technique to a new level, however, by perfecting a method of hollow casting, to produce elegant figures and ornaments with a thin wall of pure gold. They first made a model of clay mixed with powdered charcoal, and over this applied a thin sheet of pure beeswax, which was made to follow precisely the contours of the model. The pouring channel and air vents were added in the form of wax rods, attached to the wax coating. The whole was then enclosed in clay, mixed with ground charcoal. The charcoal in both inner and outer moulds was to permit gases to escape, instead of forming bubbles in the metal.

One most ingenious idea was the way the inner core was held in place, while the mould was heated to make the wax run out. This was done by a series of green wooden pegs, pushed through the outer clay casing and the wax layer, into the inner core. After the heating, and while the mould was still hot, gold was poured in to replace the wax. When the gold casting had been broken out of the outer mould, the core material inside it was extracted through a hole in the base. Finally, the holes in the gold casting where the supporting pegs had pierced the wax were plugged with gold, and polished.

The Museo del Oro in the Colombian capital, Bogota, has a dazzling display of the high artistry of its ancient gold workers. The most treasured single item consists of a raft carrying an imposing figure, guarded by four chiefs decorated with plumes and ornaments. It is an exquisite example of gold casting, and it represents one of the most bewitching tales in the whole history of man's infatuation with gold: the legend of El Dorado.

Soon after the conquistadors arrived in Colombia they had begun to hear stories from the Indians of a god-like ruler who scorned the usual finery or ornaments and went about all covered with powdered gold 'as casually as if it were powdered salt', according to one Spanish account. It was said that 'El Dorado' (the gilded man) was the central figure in a ceremony which took place on the appointment of a new ruler. His first duty was to make a sacrifice of gold to the demon god. To do this, he was stripped and his body oiled and coated with gold dust. He was then rafted to the centre of Lake Guatavita, where he threw a great pile of gold and emeralds into the water as a sacrifice.

One thing about the legend of El Dorado is real. Lake Guatavita lies in the hills of Colombia, in a deep, almost circular volcanic crater. Since the Spaniards first heard the story there have been innumerable attempts to drain the lake and collect the treasure, whose value has been wildly put as high as $300 million. Over the centuries many gold items have been dredged from the mud, but these may well have been thrown in by the pilgrims who still visit the lake to honour the legend. It now seems unlikely that the gold of El Dorado will ever be found, for in 1965 the Colombian government brought Lake Guatavita under legal protection, as part of the national heritage.

Apart from tales of blood and gold, and an empire won with the sword and the cross, many long-term effects flowed from the Spanish conquests in the New World. With its plunder, Spain was, if only temporarily, raised to new heights of power and influence. The existing supply of money in Europe was doubled in less than a hundred years, and quadrupled in two. At the same time, this new source of wealth helped to end the financial monopoly of the silver mines of central Europe, and of the German merchants. The pattern of trade and commerce was broadened, as other European powers began to send their ships and explorers around the world. Now other commodities began to play an increasingly important role in world commerce: tea, coffee, spices, silk, cotton, ivory, sugar and slaves.

Throughout all previous human history gold had been the most desired of all minerals. It had been a sinew of war and the basis of power. But the Spanish conquests of the Americas marked the high point of gold's importance as an arbiter of national aspirations. The wind-swept ruins of Machu Picchu may be regarded, therefore, as not just an epitaph to the Incas, and a world in which gold was God. We may conveniently use it to mark the beginning of a gradual decline in the once total authority of gold itself. Whatever the value still carried by precious metals – and the attachment of man himself to gold was far from ended – the time was coming when the growth and expansion of nations would no longer depend upon how much gold or silver they had.

There would be another and equally profound change in the role of gold in human affairs. From being the perquisite and possession of kings it would offer, for the first time, the opportunity of fortune to the common man. This revolutionary change was to be one of the consequences of the discovery of the New World. But it would be brought about not under the rule of any hereditary European empire, but in the new commonwealth of the people, in the United States.

CHAPTER 6
A world-wide contagion

At the beginning of the nineteenth century California was a vast, remote, neglected remnant of the old Spanish empire. Its history after the Spanish conquest had been comparatively peaceful. When devoted priests had arrived to establish a number of missions in this northern province of Mexico, the local Indians were peaceable and ready to be converted. There was none of the bitter fighting which took place between the Indians and the Spanish settlers in what is now the American South-west. There, the Comanches and Apaches waged unrelenting war in defence of their lands against the descendants of the conquistadors.

Nothing changed very much in California after Mexico became independent from Spain in 1821. The non-Indian population, mainly of Spanish descent, grew slowly to about fourteen thousand. They were scattered along the coastal plain, engaged chiefly in farming, fishing, trading, and prospecting. Little notice was taken when one of the largest townships on the northern coast, with a population of about eight hundred, decided to change its name from Yerba Buena to San Francisco.

Beyond the Rocky Mountains, however, a wave of human enterprise was steadily advancing towards California. It was led by the pioneering settlers of the new-found United States of America, who had already braved the overland journey from Missouri to occupy the Oregon country, on the Pacific coast further north. By the 1840s a few adventurous Americans had pushed on to California, and liked the look of the place. So when disputes between the United States and Mexico over the border territories escalated into war in 1847, California was swiftly annexed.

Within a year, this sleepy, sun-drenched territory, with its history of earth tremors, was to be the epicentre of a man-made disturbance whose reverberations would be felt worldwide. The hunger for gold would once again draw determined men across the world to the

The Californian gold strike in 1848 set off one of the greatest mass migrations in human history. People rushed to San Francisco by wagon across the United States, by ship around Cape Horn and, as here, by mule and on foot across the Isthmus of Panama. *California State Library*.

Americas, as it had drawn the conquistadors, three centuries before.

There was, however, a profound difference between the Californian gold rush and that single-minded plunder of gold which had animated the Spanish conquests. Like that initial beach-head established by the Old World on the New, the settlement of North America by the Pilgrim Fathers had allied faith with the hope of material gain. But it was motivated by a dream of more diverse and widespread riches. The golden summits of American institutions, like the State Capitol of California in Sacramento, capped the ideal of an individual democracy, not the authority of an absolute monarchy. That ideal stood for the commonwealth, or independent community, as well as for the idea of individual self-enrichment. That vision was to inspire a never-ending flood of immigration from the Old World, which became a limitless human resource.

However, the gold which gleams on the dome of the Capitol is also a reminder that even in the new democracy the old Adam lay in waiting, to arouse the common citizens of the nineteenth century, as he had aroused the conquistadors. For it was on the bank of the American River near Sacramento, on 28 January 1848, barely a year after California had been annexed from Mexico, that a man plunged his hand into the tailrace of a timber mill and pulled out a shining pebble. That simple action started a human avalanche.

The scene for the eventful curtain-raiser had been set by a Swiss merchant adventurer named John Augustus Sutter, a substantial figure in the territory. Sutter had arrived in California in 1839. After adopting Mexican citizenship, and receiving a twenty-five thousand hectare land grant near the American River, he prospered. He kept a large workforce of Indians and two Hawaiian mistresses, and lived in considerable comfort in an imposing complex of wooden buildings which he called 'New Helvetia', but which was known locally as 'Sutter's Fort'.

Late in 1847 Sutter went into partnership with a carpenter named James Marshall to build and operate a sawmill. Marshall decided to site the mill on the banks of the American River, since he intended to use a waterwheel as the source of power. Construction was almost finished when Marshall discovered that the tailrace below the wheel was not deep enough to take the flow of water necessary to turn the wheel, so he ordered his men to deepen it. Each night he opened the sluicegate above the mill and allowed water to flow through the tailrace, to clear away the mud and sand.

'One morning in January – it was a clear, cold morning; I shall never forget the morning – as I was taking my usual walk along the race after shutting off the water,' wrote Marshall later, 'my eye was caught with the glimpse of something shining in the bottom of the ditch. There was about a foot of water running then. I reached my hand down and picked it up; it made my heart thump, for I was certain it was gold.'[38]

Marshall hammered the pea-sized lump of metal on a rock, and when it flattened without breaking he knew he had found gold. He showed it to two of his men, and next day a few more small pieces of gold were found in the tailrace. Marshall told Sutter, who immediately went off to obtain a lease on the area around the mill from the local Indians, and then tried to obtain title to the land from the military governor of California. Word leaked out, and a few miners began prospecting along the river near Sutter's Mill. But still there was no unusual interest. Gold had been found here and there in California for decades, but no discovery had ever amounted to much.

On 14 March 1848, under the heading GOLD MINE FOUND, San Francisco's first newspaper, the *Californian*, carried a short account: 'In the newly made raceway of the Saw Mill erected by Captain Sutter on the American Fork, gold has been found in considerable quantities. One person brought thirty dollars worth to New Helvetia, gathered there in a short time. California, no doubt, is rich in mineral wealth; great chances here for scientific capitalists.'

There were, however, not many people in California to read and spread the news, and it was not until about a month later that rumour was confirmed, and the blood of prospectors began to tingle. The trigger was the sudden and unheralded appearance in San Francisco of a travel-weary, excited man named Sam Brannan, who marched through the streets of the ramshackle town beside the harbour, waving a bottle of gold dust and shouting: 'Gold! Gold! Gold on the American River!'

There was sensation in the bars and boarding houses, and a stampede out of town. Within a fortnight hardly a man was left. Crews deserted their ships in the harbour, and in some cases were followed by their captains; one left the ship in the care of his wife and daughter. A peace treaty had just been signed with Mexico, and newly-idle troops deserted in whole platoons – 716 out of a total of 1290 stationed in northern California. One deserter wrote later: 'A frenzy seized my soul; piles of gold rose up before me at every step; thousands of slaves bowed to my beck and call; myriads of fair virgins contended for my love. In short, I had a violent attack of gold fever.'

The news gradually filtered back to the eastern states, but at first was treated with scepticism. Some Congressmen were inclined to attribute the rumour to a 'Yankee invention got up to reconcile the people to a change of flag'. But as the rumour persisted, and was carried by travellers and traders departing from California to other parts of the world, there finally developed one of the most extraordinary mass migrations in history.

From every corner of the globe people rushed to California. Prospectors, miners, gamblers, instant adventurers and business-minded women made their way round Cape Horn, through the swamps and jungles of Panama, across the bandit country of Mexico, over Indian territory in the United States. In one month sixty ships sailed from New York City, and seventy each from Philadelphia and Boston. Forty-five ships from the east coast reached San Francisco in a single day. In one three-week period, eighteen thousand wagons were floated across the Missouri River at the point where Kansas City now stands. And as the eager fortune-hunters surged across the prairies in that breathless charge, in 'prairie schooners', on horseback, walking and pushing hand-carts, they opened up the West, and they speeded up the settlement of the North American continent.

In California, like a swarm of locusts, the new arrivals spread out up and down the American River and beyond, to the streams tumbling out of the foothills of the Sierras, to feed on gold. They found a hoard of unimagined richness, untouched and unclaimed. For millions of years, under the slow but insistent effects of erosion, a great 'mother lode' of gold high in the mountains had been shedding flakes, grains, and nuggets into the rivers. The heavy gold had collected in placer deposits

San Francisco harbour at the peak of the gold rush, 1849. *California State Library*.

along stream beds, in quiet backwaters, and behind rock bars across the streams. It lay in concentrations which had never been seen before. One man found a single pocket of gold dust which nearly filled his hat.

And the gold was there for the taking. It belonged to whoever found it. For the first time in history, fortunes were lying at the feet of the common man. California had no mining tradition, no prior claims by land owners, no royalties or fees to pay, no laws concerning ownership of minerals. To the homeless Irish, driven from their own country by the devastating potato famines; to the Chinese, free to emigrate for the first time in their history after the downfall of the isolationist Ching dynasty; to the unemployed Cornish miners, their industry ruined by the huge copper finds around the Great Lakes and in Australia – to men like these, the Californian gold rush offered the only opportunity that they and millions like them would ever have to better their lives.

Spurred on by individual initiative, the miners practised private enterprise in its simplest form. There were criminals among them, but also men of considerable talent and education. Their codes of conduct were crude, but they maintained a form of rough justice. The one

Early arrivals on the California goldfields had little equipment other than the wash-bowl, adapted from the *paella* pan of the Spanish settlers. *California State Library*.

inerasable blot on this generally exhilarating period was the treatment of the Indians. At first welcomed as workers, the men were driven off the goldfields in 1849 by massacres. Their women were forced into prostitution and their children into virtual slavery.

In the absence of any existing mining laws, rules had to be formulated on the spot, in the mining camps. The miners from Saxony and Cornwall introduced their long-established traditions, by which a man who first staked a claim could work it. The principle was enforced with a directness bordering on lynch law. The consensus which developed in those first hectic months on the diggings there grew into an acceptance by later state governments of the principle that rights to a mining claim and its minerals rested upon discovery, appropriation, and development. These customs were later validated by legislation, and formed the basis for the United States mining laws of 1872, most of which survive today.

That first euphoric phase of the great Californian gold rush lasted only about five years. The placer deposits eventually gave out, and with them disappeared the real opportunities for the individual fossicker. But the bonanza for America itself had only just begun.

In 1850 a prospector named George McKnight, working back into

An early development in California was the 'long tom' cradle, used to separate the gold-bearing sand from river gravel. *California State Library.*

127

The contagion of gold fever spread rapidly across the world from its source in California. With it went the tools and techniques of the 'forty-niners'. *On the dry inland goldfields of Western Australia, where there were no surface streams, the rocker cradle was adapted from washing to 'dry blowing', to separate the alluvial gold from sand and gravel.*

the mountains above the alluvial gold districts, came upon an outcrop of white quartz banded with gold, like a layer cake. It was so rich that men feared gold might lose its value. What McKnight had discovered was an exposure of the great 'mother lode', part of a mineralisation which was later found to extend for hundreds of kilometres along the spine of the high Sierras. Here was gold in even greater abundance, but it was bound tight in a matrix of quartz in the deep volcanic vents which had brought the gold from the earth's interior.

Mining gold-bearing quartz and treating it was not an activity for the lone prospector. Now it was up to the big companies. It took capital to drill shafts and drives thousands of metres into the mountains, and construct the batteries to crush the ore and extract the gold. McKnight's find became the famous Empire-Star mine, which eventually developed three hundred and twenty kilometres of underground workings. The main shaft followed the lode down into the mountain, on an incline, for more than three kilometres. In its lifetime, more than $120 million in gold came out of the Empire-Star.

The first year of the California rush yielded ten million dollars in gold, valued in nineteenth century dollars. Within five years the total had gone up to $350 million. Altogether, the young American commonwealth benefited by some two billion dollars worth of gold – greater mineral wealth than any nation in history had ever won. The fifteen million dollars paid to Mexico for California, Nevada, New Mexico and Utah turned out to be one of the most profitable land deals of all time.

On what began at Sutter's Mill the Americans erected a pyramid of capital and credit. Gold was the foundation of an economy the like of which had never been seen. The westward migrations from the populous east, and the pell-mell completion of the transcontinental railroad network, accelerated the nation-building process. But gold fever did more than finance national prosperity. It generated shock waves which over the following sixty years were felt in widely separated parts of the world.

And what of Sutter and Marshall, who began it all? Neither did well from their discovery. Sutter's property was overrun in the first mad rush. His fort and his mill were burned, and his cattle were slaughtered for food by hungry prospectors. Within four years he was made bankrupt, when the courts ruled that his land lease from the Indians was invalid. For years he tried to obtain redress from Congress, while living on a small monthly grant from the State of California. Sutter died in 1880, just when the US government was preparing to recognise his claim. Sutter's Fort, perfectly restored, now stands within the city limits of Sacramento.

Marshall fared even worse. After his first historic find he prospected other streams, but never made another strike. He lived for a while on a small pension from the State government, but died in poverty at the age of seventy-four, an embittered man. His only reward was the monument belatedly erected over his grave, on a hill above the site of Sutter's Mill (which is also restored, beside the American River). Marshall's tall, gaunt figure in bronze points to the place below where he ignited that great conflagration of human hopes and desires.

History has recorded a rather happier fate for the man who carried the contagion of gold fever across the Pacific to Australia, in 1851. His name was Edward Hammond Hargraves.

Hargraves was born in England in 1816, and at the age of sixteen made his way to Australia as a sailor. There is little information about his early life in the colony of New South Wales, except that it covered a lot of territory and many occupations. At one time or another Hargraves was a station overseer, a property manager, a shipping agent, a hotel proprietor and a cattle raiser. He also developed a glib tongue and an air of knowledge which many found convincing. By 1849, when he heard of the Californian gold rush, Hargraves was thirty-two, a man of 'huge physique and slim principles'. With boundless confidence he sailed for San Francisco to find gold, leaving his wife and five children with promises of imminent fortune.

As it happened, Hargraves found no gold in California, but when he returned to Australia he did become involved in the discovery of payable gold in New South Wales. As a result, Hargraves has gone into the history books as Australia's principal benefactor in this field, the 'father' of the gold rushes which were to have such a profound effect on Australia's social and economic development.

There are, however, two quite different versions of this important chapter of Australia's history, and Hargraves' role in it. One is based largely on Hargraves' own accounts after the event, both published under his own name (through a ghost-writer) and given in evidence to various official enquiries. Such was his prowess as a professional persuader that his accounts have crept into many published histories, and have certainly shaped the public image of Edward Hargraves as 'the man who discovered gold in Australia'. Another version altogether has been compiled by the Australian historian, Geoffrey Blainey, from a study of contemporary documents, government papers, and the proceedings of the various enquiries after the rush was over.[39]

According to Hargraves' own accounts, he spent two years in California, on the American River and in the foothills of the Sierras. He records a visit which he is supposed to have made to Sutter's Mill in the summer of 1849 to buy timber, and a conversation that he had with the famous James Marshall. When Hargraves mentioned that he was from Australia, Marshall is reported to have said: 'Then why don't you go and dig among your own mountains? From what I have heard of that country, I have no doubt whatever that you would find plenty of it there.'

According to Hargraves, that very possibility had already occurred to him. He recalled later that the geology and landscape of California had reminded him irresistibly of country he had once visited near Bathurst in New South Wales. During the winter of 1849, while sheltering in lodgings in San Francisco, from the cold up in the diggings, Hargraves wrote to a friend in Sydney: 'I am very forcibly impressed that I have been in a gold region in New South Wales, within three hundred miles of Sydney'. He had dug much gold, he wrote later in his book, 'but the greater our success was, the more anxious did I become to put my own persuasion to the test, of the existence of gold in New South Wales.' And so he returned to Sydney, 'to find payable gold'.

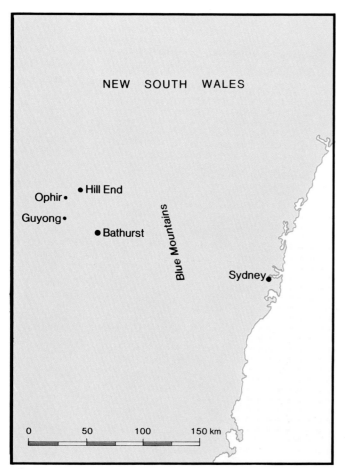

The valley of Lewis Ponds Creek near Guyong in NSW, where Edward Hargraves found traces of gold in 1851, and set off Australia's first gold rush. *The shallow depressions in the bank are where those who followed Hargraves dug hopefully into the beds of alluvial gravel.*

The facts, such as are known, differ somewhat. Hargraves spent only a year in California, and found no gold whatever there. He seems to have spent most of his time wandering about the goldfields, chatting to the 'forty-niners' and exercising his undoubted talents as a raconteur and publicity agent for himself. His decision to return to Australia was influenced as much by his desire to avoid another freezing winter in the Sierras, and the pleas of his wife, who was struggling to exist with their five children near Gosford, as by his visions of gold in New South Wales.

However, by the time Hargraves' ship reached Sydney his confidence was restored and brimming over. He was now 'the man from Sacramento', a large and commanding figure, and when he announced to anyone who would listen that he was off to find gold, there were many who were impressed. One who was not was the Colonial Secretary, Deas Thomson. Hargraves had persuaded a solicitor friend to write to the government requesting a grant to finance the expedition. Thomson refused. But Hargraves managed to borrow enough money from a merchant to buy a horse and stores, and on 5 February 1851 he rode west out of Sydney and over the Blue Mountains.

Hargraves was not just relying on his much publicised recollection of country that reminded him of the gold-bearing hills of California. In fact, his claim to knowledge of geology seems quite suspect, despite his assertion in a Sydney newspaper article that 'I myself, sir, have never yet been deceived by nature'. However, Hargraves was well aware that gold had been found in many places in the Bathurst-Wellington area, although not in payable amounts. It had invariably been in the form of isolated nuggets or outcrops, stumbled over by shepherds, who had ample time to study the country as they walked across it.

The rocker cradle was another device used by the 'Forty-Niners' to separate gold-bearing sand from gravel, before finishing the washing process with the bowl. *California State Library*.

One shepherd, named McGregor, was known as the 'gold-finder', because as early as 1845 he had taken small lumps of gold into Sydney for sale to jewellers. Like other shepherds, McGregor had chipped his gold out of quartz outcrops. Neither he nor any of the others had any conception of the existence of alluvial gold in the river sands, or of the techniques of recovering it. And that, perhaps, is where Hargraves felt he could succeed. The only thing of value he had brought back from California with him was the knowledge of how to use the washbowl and the wooden rocking cradle, which had been the primary tools of the 'forty-niners'.

From Hargraves' own account, it seems that on that initial excursion he was heading for Wellington, the area where McGregor the shepherd had found gold. But on the night of 10 February Hargraves stayed at a hotel in a little place called Guyong, north-west of Bathurst. The inn was on the edge of a highly mineralised stretch of country, where copper had been mined, and many small finds of gold had been made. Hargraves said later that 'almost every mantelpiece was crowded with ore'.

In the bar that night there was talk of the recent find of gold that 'Yorky the shepherd' had made not far away, on Lewis Ponds Creek. Hargraves immediately declared that he knew the area well, having once travelled over it in search of lost bullocks, and announced his intention of prospecting up that way. In fact, there is no reason to believe that Hargraves had ever been anywhere near that area.

Hargraves borrowed a tin dish, a pick and a trowel, and on the morning of 12 February, with the innkeeper's son, John Lister, as a

guide, he rode north along the dry valley of Lewis Ponds Creek. It was the height of summer, and there was drought in New South Wales that year, so the creek had become an intermittent string of waterholes. About noon Hargraves and Lister stopped to make tea near a gravelly stretch of creek, with a little water in it. Then, like Joan of Arc, who claimed to 'hear voices', Hargraves announced that he felt 'surrounded by gold'.

Watched no doubt with fascination by young Lister, Hargraves took his tin dish and went down to the water's edge. He washed a panful of gravel, and in it found a speck of gold. He washed five more panfuls, and in all but one found gold. To be true, no wash had yielded more than a single tiny grain of gold, but it was enough for Hargraves. As he wrote later, he immediately made a speech to his audience of one: 'This is a memorable day in the history of New South Wales. I shall be a baron, you will be knighted, and my old horse will be stuffed, put in a glass case, and sent to the British Museum.'

Back at the hotel that night, Hargraves found his audience somewhat less impressed with his finds, which he showed around in a folded sheet of paper. One visitor could see the specks of gold only with a magnifying glass. Another peered at them through the base of a glass tumbler. Next day Hargraves went out again, this time with a second companion, a young drover named James Tom. While Hargraves rode along the stream banks on his horse, issuing instructions, the other two panned ceaselessly for gold. They found specks, but no more. Hargraves then showed Lister and Tom and Tom's two brothers how to make and use the Californian rocker cradle, which enabled much

Above left: Hargraves brought back to Australia the knowledge of the two characteristic tools of the 'Forty-Niners', and showed eager prospectors how to use them. *Above*: If the miner was lucky he was rewarded with a few grains of gold in the pan.

133

more material to be treated. But although they hauled the cradle along the creeks for days, they still did not find gold in payable amounts.

In March Hargraves rode back over the Blue Mountains to Sydney, while the others continued the search around Bathurst and the Macquarie River. He had only a few grains of gold to show for his expedition, but this did not particularly worry him. For Edward Hammond Hargraves, during that ride, may well have begun to formulate the plan which was to bring him fame, comparative comfort for the remainder of his life, and an enduring place in the history books.

From his visit to California, Hargraves undoubtedly knew that looking for gold was a gigantic lottery, and that his chances of winning it were slight. He was neither suited to nor practised at the art of panning alluvial gold. He was also undoubtedly aware of a very real risk – that anyone finding gold in New South Wales might not be able to keep it. Unlike California, which had no laws concerning minerals, there existed in Australia the unwritten but implicit right of the Crown to the land and all that went with it. And so, it seems, Hargraves decided that his best chance of benefiting from any discoveries of gold would be in the form of rewards from the government, not from prospecting. All his subsequent actions support this view.

Hargraves must have realised that for him to appear as a public benefactor he would need popular support. So he set out to arouse excitement, of the kind he had seen in California. He found a receptive population. In that year of 1851, because of the drought, the colony of New South Wales was in depression. Wool prices were down, the boom in land sales had slumped, and in general the outlook was gloomy for the scattered settlers still clearing their land and trying to survive. They, and the colony as a whole, were in the mood for some kind of deliverance. Hargraves resolved to help them find it.

His first move, on reaching Sydney after a long ride through the rain, was to go straight to see the Colonial Secretary, Deas Thomson. After being kept waiting for three hours, Hargraves confronted Thomson in his still soaking overcoat and declared that he had found a valuable goldfield. He asked for money to cover the cost of his expedition, and said that he would 'await the generosity of the government' for the rest of his reward. When asked to support his claim he produced his piece of folded paper containing a few grains of gold. Thomson could hardly see them, and sent Hargraves away, unsatisfied.

Hargraves then persuaded a friend to write a letter to the *Sydney Morning Herald*, reporting the finding of a large goldfield by Mr Edward Hargraves, and urging the government to 'reap the golden harvest'. This publicity provoked the still sceptical Colonial Secretary to write to Hargraves, asking him to provide more precise information about his 'goldfield', and deferring consideration of any reward until the newly arrived government geologist had inspected it. Hargraves was about to reply when news arrived from Bathurst which dramatically supported his story: John Lister and William Tom, two of his novitiate prospectors, had found four ounces of gold at 'Yorky's Corner'.

Wasting no time, Hargraves wrote two letters to Thomson. In one he confidently described Yorky's Corner (although he had never seen it) and the area that he had unsuccessfully worked. In the other he asked for a loan of thirty pounds to buy a new horse, so that he could return to the scene and show the government geologist around. The Colonial Secretary reluctantly agreed. Hargraves rode off again over the Blue Mountains, to set his plan in motion.

Hargraves had apparently decided to take a bold gamble, totally at variance with traditional prospecting practice. Instead of concealing as long as possible the whereabouts of the gold finds, he would broadcast the secret to the world. The more people there were looking for gold, the better the chance that it would be found, and the more pressure there would be on the government to reward him. So Hargraves went back to the inn at Guyong, bought the four ounces of gold that his friends had found, and sent some of it to the Colonial Secretary with the clear implication that he had found it himself. Taken by his friends to the place where the gold had been found (not far from the present township of Hill End), Hargraves grandiloquently named the district Ophir, after the Old Testament 'place of gold'.

Hargraves then embarked upon a masterful campaign of publicity and propaganda. He gave public lectures in Bathurst, displaying samples of gold, and telling his eager listeners where to find more. He went down to the Ophir district and showed new arrivals how to use the pan and rocker cradle. He went to meet the government geologist and, like the Pied Piper, led that official and a growing horde of diggers back to Ophir. Within days there were hundreds of men frantically digging and washing in the streams. 'The effect of my appearance in the district has caused a little excitement amongst the people,' wrote Hargraves with careless modesty to the Colonial Secretary.

To Hargraves', and Australia's, vast good fortune, his prophecies turned out to be true, and beyond all reasonable imaginings. Just as in California, amazingly rich pockets of alluvial gold were found in those ancient valleys and river beds, washed down from lodes back in the mountains. From all over the colony, people rushed to the scene. By the end of May the governor of the colony, Fitzroy, was writing urgent despatches home to London reporting 'the very great excitement which is engrossing and unhinging the minds of all classes in the community'.

As for Hargraves' persistent claim to have been the 'discoverer of gold in Australia', it was not sustained by commissions of enquiry at the time, or by objective historical studies, except in one respect – it was undoubtedly Hargraves who engineered the first Australian gold rush, and he did it virtually single-handed.

Compared to Sutter or Marshall in California, Hargraves received quite substantial rewards. In May, 1981, the government, worried that the Ophir field might give out, and the 'vast herd' of prospectors have nowhere else to dig, made Hargraves a Commissioner of Crown Lands, and asked him to find new goldfields. Hargraves reassured the Executive Council that 'the southern parts of this Colony, Goulburn, Gundagai, and Murrumbidgee, from what I have heard, [would be] more rich than the western', and he went off with a servant and pack-horses to search.

Needless to say, Hargraves never found another Ophir, although he rode unknowingly across many gold-bearing areas that were later exploited. However, he eventually received cash rewards from the government of ten thousand pounds and a handsome pension, and by public subscription an assortment of gold cups, spurs, whips and nuggets – worth a quarter of a million dollars by today's standards.

In one way, however, Edward Hargraves may have had a much more far-reaching influence on Australia's history than even he had imagined. This concerns his role in the framing of Australia's mining laws, with all their political consequences. 'In 1851,' writes Blainey, 'through the success of Hargraves in creating a rush weeks before a rush would have normally begun, and through the doubts of the government that its territory was rich in gold, a crisis had emerged. It was a far graver crisis than Australia faced in 1914 or 1939 because it was so sudden, so novel, so revolutionary in its social implications with its fears of famine and bloodshed.'[40]

The crisis had arisen from the fact that the colony of New South Wales had no mining laws. The governor and his officials had no idea how they could or should control the rush for gold that had suddenly developed. They thought that the gold belonged to the Crown, but they could not be sure. They were quite sure that it did not belong to the motley mob that was busy digging it up, but they did not know how to stop them. The Commissioner of Crown Lands on the spot wrote hopelessly that 'I have served many of the numerous parties now digging for Gold with Notices to desist, to which notices they pay but little attention.' There were, after all, only about three hundred troopers in the entire colony. 'It would be madness to attempt to stop that which we have not physical force to put down,' the Colonial Secretary told James Macarthur and other pastoralists, who were calling for martial law against what they perceived as a threat to their tenure.

There was another aspect to the crisis. The governor, Fitzroy , feared that shepherds and drovers would leave their sheep, although shearing time was near; that farmers would leave their ploughs, thus threatening the population with starvation; that policemen would desert their beats and jailers their prisons, inviting widespread lawlessness; that shopkeepers would abandon their businesses, teachers their schools, sailors their ships. The life of the colony might, in fact, be totally submerged in the rush for gold.

There was no precedent, no source of guidance for Fitzroy and Thomson. Neither England herself nor any of the colonies had ever had to face such a situation. The Californian example was both unhelpful and disturbing. Mexican law had ceased, and American law had still to be imposed. The main message received by the officials in Sydney was the danger of mobs that craved for gold. In the end, however, the government was forced to act. Fitzroy invoked a sixteenth century lawsuit in England, *Queen v. Earl of Northumberland*, and proclaimed the Crown's right to all mines of gold in New South Wales, and threatened to prosecute all who sought or dug for gold without permission. Such permission would only be conveyed by a licence, issued by the government.

136

The discovery of gold in Victoria later in 1851 started an even bigger rush, as men everywhere left their jobs and headed for the goldfields. *Mines Department of Victoria.*

In this decision, it now appears, the government of New South Wales was acting under the impression that it was following Californian practice where, it believed, a similar tax had been imposed on miners. In fact, California imposed no such tax on its citizens, but actively encouraged people to look for gold, and to keep it when they found it. Who, then, could have misled or misinformed the government? In Blainey's view, 'the only available evidence points to Hargraves.' Hargraves himself later claimed to have influenced the framing of the regulations, presumably because the more money the government raised from licence fees the greater his chances of gaining high reward. Whatever the reasons, the decision by the colonial government to make gold mining subject to a licence fee was to have profound repercussions.

Initially, the licensing system worked quite well, chiefly because of the enlightened attitude of the man who had to enforce it: John Hardy, the new Commissioner of Crown Lands for the gold district. With only ten men he had to issue the licences, administer the diggings, settle quarrels and administer justice. In Blainey's description, 'Hardy reached Ophir on 3 June and saw over a thousand men camped along the creek. He had no tent, no table, not even a candle by whose light he could write at night the reports that the government nervously awaited. On his first morning he went down the creek and collected the tax with so little trouble that he complained only of the tedium of repeatedly writing out licences, weighing the gold that the diggers offered in payment, and marking out the area to which he decreed each man was entitled.'[41]

137

Hardy exercised his powers with understanding. The fee set by the government – thirty shillings a month – was quite high for the colony, where the average station hand got about two pounds a month, plus keep. Hardy let men work for a few days without a licence, to see if they could find enough gold to pay for it. Thousands came, searched, and left without having to buy a licence.

But before long the government's attitude hardened. After all, the idea of the licence was to force unsuccessful diggers back to their usual work in the colony. The licence itself was hedged with restrictions – no man could get one without the permission of his previous employer – and it carried a description of its holder, so that it could not be transferred. As Blainey sums it up: 'As a crisis measure the digger's licence was masterly in design and initial execution. The tragedy was that like many taxes it remained long after the crisis had passed. In 1854 it was to incite an even graver crisis – armed rebellion at Ballarat.'[42]

The events which led up to the Eureka Stockade, Australia's only civil uprising, are beyond the scope of our story, except for the fact that the gold fever which struck Victoria later in 1851 completely over-shadowed the rush which Hargraves had set off in New South Wales. The discovery of enormously rich alluvial gold around Ballarat and Bendigo almost brought Australia to a standstill, as people everywhere abandoned their occupations and flocked to Victoria. Then, as with California, the overseas fortune-hunters began to pour in. The population of Victoria doubled in a year, that of Australia trebled in a decade.

Of course, for every one who struck it rich there were thousands who failed. Without the resources to go back where they had come from, they were forced to turn to other enterprises. The impact of this sudden injection of human resources into the colonies was overwhelming. It produced the greatest social and economic changes in Australia's short history. The country had, after all, been settled by Europeans for only about sixty years. Primary and secondary industries boomed, and railways began to spread out from the cities to the country towns, to bring in the growing harvests of wool and wheat. In a political sense, the influx of free spirits swamped the ex-convict element in the population. Workers began to provide a balance to the two existing blocs of power, the administration and the pastoralists.

As in California and New South Wales, the great Victorian gold rush was at first based on the recovery of alluvial gold from streams and dry river beds. Most of this gold was in the form of dust or grains, and had to be washed from the sand and gravel. But there were also some astonishing finds of alluvial gold which occurred as huge lumps or sheets, often lying close to the surface. One such mass, named the 'Welcome Stranger', yielded 2284 ounces (84.5 kilograms) of pure gold when it was melted down, and is probably the largest single piece of gold ever found.

Before long, however, the diggers began to follow the gold below the surface. They had discovered Victoria's unusual gold-bearing for-mations: the remarkable system of 'deep leads', or buried rivers of gold. These were formed long ago, by the erosion of gold from 'mother lodes' in the mountain ranges into the stream beds, in the familiar

When the alluvial gold gave out the miners had to sink shafts to seek the 'deep leads', and the horse whim came into use to raise the ore. *This is one of the working displays at the Sovereign Hill Gold Museum at Ballarat, Victoria.*

process of alluvial deposition. Then an accident of geological history buried these gold-laden rivers beneath a vast flow of material which poured out of a chain of volcanoes up and down eastern Australia, tens of millions of years ago. This huge blanket of basalt, which covers most of Victoria's Western District, was itself slowly cut by streams, in which alluvial gold again began collecting. This was the gold which sustained the first great Victorian gold rush. But below the present stream beds lay those buried rivers. They had flowed far longer than the existing streams, and had collected much more gold, in places in the form of massive nuggets. And there, in those water-logged 'deep leads', the gold lay waiting to be found.

The problem in mining the 'deep leads' was that there was no known technique for locating the gold, no geological guides to follow. The existing surface topography bears no relation to the ancient landscape and its winding river beds. Finding the buried watercourses became one of the biggest gambles in the history of gold mining. Eventually, it was left to large companies to tackle on the scale required.

Around Ballarat, the fossickers' technology did not go far beyond sinking a shaft, with a windlass at the top to haul up the spoil. As they went down they invariably struck water, which they had to bale out with buckets. Loose gravel poured in on them. Sometimes the shaft itself collapsed around them. In the heat of summer they nearly suffocated at the bottom of the airless shafts. And when they reached the deep river beds they did not always find gold. Some who did made fortunes.

By 1854 there were hundreds, perhaps thousands of such mines in Victoria. The average shaft was about fifty metres deep, and took five to eight months to sink. Most of the miners got nothing out of them, after working for half a year or more. And all this time they had to pay their monthly fee for their hated licence. They were frequently harassed, too, by officious policemen, who could oblige a miner working below to leave his work and climb up the shaft to produce his licence, then go all the way back down again. The inequities of the licensing system were bad enough on the alluvial fields, but they worked particularly harshly on the deep miners around Ballarat. These were contributing factors to the rebellion against the whole system in December 1854, which saw a thousand men barricade themselves in their stockade on the outskirts of Ballarat and defy the troopers.

The dawn attack on the Eureka Stockade was fierce but brief. Five soldiers and about thirty miners were killed. The effects were to be much more protracted. 'In Australia's quiet history,' writes Geoffrey Blainey, 'Eureka became a legend, a battlecry for nationalists, republicans, liberals, radicals, or communists, each creed finding in the rebellion the lessons they liked to see.'

The most immediate result was the abolition of the mining licence, and its replacement with the 'miner's right', which for a pound a year entitled a man to dig for gold, vote at parliamentary elections, and make his own mining laws. The Californian system at last came to Australia. The miners elected the members of the mining courts, which made all local mining regulations, decided how much ground each man could hold, and on what conditions he could retain it. 'This system was

The influx of free spirits into Australia during the gold rushes swamped the convict background of the colonies, and transformed the nature of Australian society. *La Trobe Library, Melbourne.*

probably the high tide of Australian democracy,' concludes Blainey. 'The miner in moleskins, for long hunted and herded, was lord of his own goldfields.'

The finding of gold in Australia, as in California, had a profound effect on the nation's economy, and would do so in other parts of the world where gold was soon to be discovered: New Zealand, South Africa, and Alaska. The gold rushes, wherever they occurred, brought new settlers, new ideas, new vigour, and created new wealth. Without the enormous amounts of gold that were produced in the latter half of the nineteenth century the commerce of the modern world could never have reached the proportions that it has today. Only after the gold rushes was it possible to speak of something called world trade.

Gold was valued because, in general, nations and businesses and even individuals did not really trust one another. They placed little faith in paper money, promissory notes, cheques or contracts – unless they were backed by something which they all trusted, and that was gold. Gold became the standard of value, the symbol of trust between men and nations. As for Australia itself, had it not been for the gold rushes, it might well have remained for much longer in its history a convict settlement, a place of exile and banishment.

There was one group of miners whose role in the golden era of Australian history was different from all the others. Although at first they had nothing to do with gold mining, they rapidly became indispensable to it. In doing so they helped to link what was then still a remote colony with the world-wide community of underground miners, which since Neolithic times has provided the invisible labour on which the whole history of metals and metallurgy has depended. That remarkable clan was made up of Cornishmen, known on four continents as 'Cousin Jacks'.

At the beginning of the nineteenth century Cornwall was the world centre of mining. The copper and tin dug from its ancient granite formations, and smelted in Wales, were vital to the growing industrialisation of Europe. The men who worked in the deep mines were the most accomplished hard-rock miners to be found anywhere, and perhaps the strongest. The mines of Cornwall had been worked since Roman times, and the production faces lay at the remote ends of deep drives, some extending far out beneath the sea. The Cornishmen were accustomed to climbing down and up a thousand rungs of ladders, and walking as much as eight kilometres underground, in a day's work.

Their system of mining, based on 'tribute', was perhaps unique in maintaining both production and morale in an industry historically riven by conflict. By this system the mine managers, called captains, marked out the various areas down the mine, and fixed for each one a price per ton for the ore recovered from it. The miners bid for the areas they considered would be most rewarding, and worked them for two months, fixing their own hours and working arrangements. They received no wages, but at the end of the period they received the agreed percentage of the value of the ore they had brought up. The 'tributers' among the miners, who did the selection of the places to work, had to become, in effect, practising geologists. They acquired an unrivalled knowledge of ore bodies and ways of mining them most efficiently.

It was this knowledge and experience which made Cornish miners welcome everywhere they went, as they began moving out across the world in the first decades of the nineteenth century to the new lands and new opportunities in North America and the southern hemisphere. Some went to South Australia, the last of the colonies to be settled on that vast continent. While the Cornishmen were undoubtedly attracted by the physical, civil and religious freedom that Australia offered, in contrast to the rigid society they had left behind, they were also hoping to try their luck in a country which had no mining tradition, and in fact no metal mines of any kind. The only mining then was for coal, on the east coast, and on a very small scale. In Australia the Cornishmen found a mineral treasure-house, which by extraordinary chance had lain waiting, untouched, for millions of years.

For hundreds of millions of years the surface of the Australian continent had been quietly wearing down, exposing mineral deposits which in many cases had weathered and oxidised into easily recovered and processed ores. Elsewhere in the world, these mineral riches on the surface were the first to be used by man, and no longer exist. In Australia, as we have noted, the Aborigines virtually ignored them. And so the Cornishmen who arrived in South Australia, accustomed to delving deep into the earth for metals, found a landscape which they could never have imagined existed, with outcrops of metals so rich that they could not believe it.

The first discovery came in 1841, within an hour's walk of Adelaide. Two Cornish miners saw a shiny, metallic-looking rock jutting out of a grassy hillside, and recognised it as galena, the most important ore of lead. The mine which they opened up – the first metal mine in Australia – was called Wheal Gawler, 'wheal' being Cornish for mine. Soon the hillside at Glen Osmond was dotted with 'wheals', following the veins of rich silver-lead, some assaying seventy per cent lead. But the real prizes were still to come.

Late in 1842, on a sheep run sixty kilometres north of Adelaide, a pastoralist named Dutton was riding through a rain storm in search of sheep, when he saw on a hillside a patch of vivid green, more intense than anything in his experience. The rock, he said later, seemed to be covered with 'beautiful green moss'. It was malachite, the ore of copper that has been prized throughout human history, the subject of man's earliest experiments with smelting. At Kapunda it was bursting out of the hillside, so soft and rich, assaying twenty-three per cent copper, that in a few minutes a man could fill a wheelbarrow with it.

That was the beginning of an astonishing period of copper mining in South Australia. Extraordinary lodes of copper were found in the Flinders Ranges. One, at a place called Burra, was described as a huge 'bubble' of copper bulging from the earth. The Burra mine rapidly became the most important employer of labour and earner of income in the colony. The copper ore shipped to Wales for smelting exceeded in value the combined exports of wool and wheat. Burra soon had a population of five thousand people, and the backbone and moving spirits of the community were the Cornish miners and their families.

The Cornishmen turned this distant corner of the world into a little Cornwall. They built their distinctive granite engine houses over the mine shafts, and their stone cottages in the townships. They introduced the tribute system, and called their managers captains. They sang in their Wesleyan chapels, and their wives baked them pasties to take down the mines. They ran whippets and, until they were stopped, staged cock-fights.

Ironically, the skills of the Cornishmen and the copper they shipped to England helped to bring about the downfall of their own once great industry at home. The flood of easily recoverable copper from Australia and from vast finds of almost pure metal around the Great Lakes in North America helped depress the market, and forced closure on many of the older mines in Cornwall. This accelerated the movement of Cornishmen away from their homeland, to places where their skills were needed.

Then, in 1851, the South Australian copper mines were swept by rumours from the east. Gold! The news came at a critical time for places like Burra. Copper grades were falling, and the cost of shipment to England was becoming crippling. Like everyone else, the Cornish miners felt the ancient, mystical tug. And so there began a rush to the goldfields. It began quietly but rapidly swelled to a stampede. The mine captains withheld the tribute due, to try to keep their men. At one time, it was reported, so many were hurrying east on stolen horses that extra police were posted at the river crossings on the Murray to stop them.

In the search for gold, Victorian mining companies sank shafts to unprecedented depths, following quartz reefs thousands of metres below the surface. This was the Queen Victoria Quartz Company mine in 1860, at a time when Australian mining technology was leading the world. *National Library of Australia.*

In the end, few Cornishmen made their fortune on the diggings. They were not particularly knowledgeable about finding alluvial gold, or skilled at recovering it. Most of them went back to South Australia, where they remained a human resource-in-waiting for a challenge that would soon develop, to take them once more, this time permanently, from their 'Little Cornwalls' in the hills.

First, however, there was a sudden revival of the South Australian copper industry in the 1860s, with the discovery of its greatest ore body, on the Yorke Peninsula. For a few years, the mines at Wallaroo and Moonta displaced Cornwall itself as the largest copper-producing centre in the British Empire. By a fortunate coincidence, copper prices soared in the early 1870s. The copper towns held around twenty thousand people, or two per cent of Australia's total population. One mine in Wallaroo employed up to fourteen hundred men and boys.

The boom did not last for long. By the 1880s world copper prices had crashed again. The famous mines at Burra and Kapunda fell silent. At Moonta and Wallaroo they were on short time. The Cornishmen knew they had to move on – and move they did, out into the other parts of Australia, to provide the very infusion of skill and experience that was needed, as everywhere across the continent the great lodes of minerals were beginning to be recognised.

The day of the big mining company had arrived. There was still gold to be had in Victoria, but now it had to be won from the hard quartz structures deep below the surface. It needed capital, equipment, and, above all, skill at a kind of mining which had not been seen on the goldfields before. The developments which followed put Australia ahead of the rest of the world in mining technology.

Shafts were sunk to unheard-of depths – 1000, 1500, 2000 metres. Across Victoria towering poppet heads rose over the deep mines, the ore came up in growing streams, and the days and nights throbbed with the pounding of the stamper batteries. Gold production reached new levels, and dividends strengthened the economy against crippling droughts, as federation and independence under the Crown drew near. By the turn of the century those deep mines had yielded more than half the entire mineral wealth produced in Australia in the nineteenth century.

But all over the world, by this time, the men who went down into the earth for metals had come up against the same challenge that the Romans had faced, and failed to conquer. The time had come at last for the ancient arts of mining to be industrialised. Even the incomparable skills and dedication of the Cornish miners had reached their limits. The task was now beyond the power of wind or water or living muscle. Henceforth there would have to be tireless machines to pump out the ever-deepening shafts, drive the ever-lengthening tunnels, and raise and crush the ever-mounting tonnages of rock.

From mines all over the world the call went out for new machines, and a new and superhuman source of energy. It was answered from England. There, the Industrial Revolution was thrusting aside the puny endeavours of men and muscle, and changing forever the way that man would live and work.

144

CHAPTER 7
The revolution of necessity

The map of medieval England is dotted with famous fields, scenes of victories and defeats which shaped the history of Europe and perhaps of the world. But none had a more widespread or lasting influence than the events which took place on nameless battlefields of the last two centuries, scattered across the Midlands and the North, with their crumbling fortresses and their rusting arms.

They were the battlefields of a revolution led by a relative handful of resourceful, determined men, whose fortresses were the factories and iron foundries, and whose weapons were the machines and tools which helped them fashion the Industrial Revolution. Under the smokescreen of their belching blast furnaces, the captains of industry – the ironmasters and engineers – invaded the green vales and forests, and they transformed that idealised portrait of 'Merrie England' into the world's workshop. Their victory left few lives untouched anywhere, either then or since and more than two centuries later we are still trying to digest and deal with the consequences.

One of the enduring questions about the Industrial Revolution is why it should have begun in England, when the conditions which brought it into existence were also to be found in many other countries in western Europe. It was, however, the British who in the middle of the eighteenth century began a spectacular advance from their bucolic past. They generated a self-sustaining wave of vigour that within a hundred years had won them the greatest empire ever seen, and made Britain the dominant world power.

Regardless of any political or intellectual contributions to that worldwide upheaval of traditional ways of life, it is now clear that Britain provided two of the most vital practical impulses: a method of producing iron in virtually unlimited amounts, and the world's first source of energy other than the natural forces of wind, water and living muscle — the steam engine. Ironically, both these advances were brought about by necessity. They grew from the constraints imposed on the

The battlefields of the Industrial Revolution are dotted across the once 'green and pleasant land' which they converted into the world's workshop. *Chatterley Whitfield colliery in Staffordshire, once one of Britain's most productive coal mines, is now a mining museum.*

The demand for charcoal by the iron-making industry in the Middle Ages led to a decimation of the forests, and brought on Europe's first great 'energy crisis'. *One of the last of the charcoal-burners at work in a valley in southern Austria.*

countries of Europe by a problem which has returned to find a place in our own vocabulary: an energy crisis.

Both cause and consequence of that first European energy crisis can still be seen in the quiet valleys of southern Austria, where one or two charcoal burners are still practising their craft. As their predecessors have done for millennia, they stack their timber into piles and by a simple but subtle process turn it into a clean, smokeless fuel.

Charcoal making is a practice whose origins go back to man's first

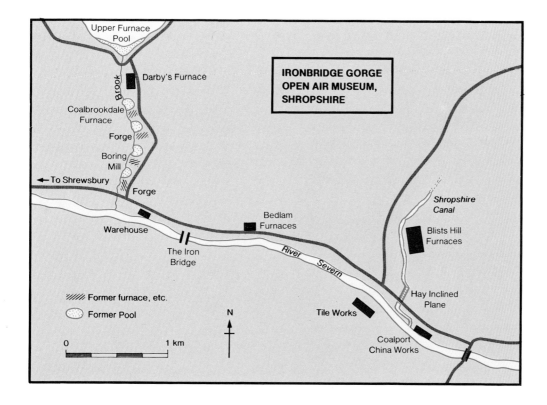

Upper Furnace Pool

Darby's Furnace

Brook

IRONBRIDGE GORGE OPEN AIR MUSEUM, SHROPSHIRE

Coalbrookdale Furnace

Forge

Boring Mill

← To Shrewsbury

Forge

Warehouse

The Iron Bridge

Bedlam Furnaces

River Severn

Shropshire Canal

Blists Hill Furnaces

Hay Inclined Plane

Tile Works

Coalport China Works

▨ **Former furnace, etc.**

◯ **Former Pool**

0   1 km

N

experience with fire, and the observation that partly-burned wood makes a more versatile fuel than wood itself. The slow, controlled combustion in the confined atmosphere of the charcoal stack drives off water and volatile elements, leaving an almost pure carbon which burns with a hot, clean glow. That discovery began one of mankind's more devastating assaults on the environment: the cutting of trees to make charcoal.

At first, no doubt, charcoal was used almost exclusively for cooking and heating, as people began to live in enclosed spaces, such as caves and huts. But when metals began to come into daily use the consumption of charcoal must have increased beyond all expectation. For the melting and working of native metals, such as gold and copper, charcoal was ideal, since it provided a source of intense heat even in a small crucible. For the large-scale smelting of copper, lead, silver, iron and other metals from their ores, charcoal was essential.

Across the ancient world, as the Bronze Age got under way, the forests around every mining area were decimated. The results can still be seen along all the shores of Mediterranean, where stands of oak and cypress once came right down to the sea. By the end of the Roman era the slopes were as bare and rocky as they are today. In some extreme cases, such as the copper-rich island of Cyprus, three thousand years of smelting saw virtually the entire island cleared of its trees to fuel the smelters.

After the stagnation of the Dark Ages the forests of western Europe came under even more sustained attack. The renewal of economic activity, the growth of populations in towns and cities, and, finally, the steady rise in the output of iron – all these created demands for energy that could only be met by the making of charcoal. At the same time, there was an expansion of farming to produce food for the growing populations, and a corresponding contraction in the areas left under forest.

A glimpse of what Europe must have been like at this time, on the threshold of the Industrial Revolution, can still be obtained in the mountains of Styria, in southern Austria. This major iron and steel-producing region is centred on the huge 'mountain of iron', the Erzeberg, which for hundreds of years has been carved away in terraces, like a monstrous layer cake. In the quiet valleys nearby there are scenes of pre-industrial life which elsewhere in Europe have for the most part been swallowed up by the growth of cities.

The little village of Vordenberg, for example, is dominated by a solid yellow brick and stone building which contains the last of a succession of fourteen iron furnaces which worked in the valley from medieval times until the beginning of the present century. In its general design the Vordenberg furnace had its counterparts all over Europe. It was fed from the top with iron ore and and charcoal, and it needed a huge amount of charcoal to smelt a single tonne of iron. The method was profligate. It was the appetites of these furnaces which devoured the forests on the once densely timbered slopes around Vordenberg, and turned them into alpine meadows.

These effects were felt all over Europe, but the first country to reach crisis point over the shortage of wood and charcoal was Britain, for two main reasons. One was the limited area of the British Isles themselves and of their forests. The other was the proportionately more rapid growth of population in Britain than anywhere else in Europe. In England and Wales, between 1530 and 1690, the population nearly doubled, from three million to just under six million. There was also a widespread move into cities, especially London, whose population multiplied eight times over that same period, from sixty thousand to more than half a million, which made it the largest city in Europe and perhaps in the world. Larger towns and cities meant heavier consumption of wood from surrounding areas, and soaring costs for fuel and timber for construction. During the Elizabethan period in England wood prices rose very much faster than those of any other commodity.

The wood shortage was exacerbated by the growing demands of the iron-making industry. This problem became acute after the development, in the fifteenth century, of the means of turning out cast iron in quantity. This was achieved by the harnessing of waterwheels to large bellows, to increase the supply of air to the furnaces. The earliest known water-powered blast furnace in Europe began operating at Ferriere in Italy in 1463. From that time onwards the new technology spread rapidly, especially when it was found that cannons could be cast from iron instead of bronze, which was expensive. At the same time, tools, agricultural implements, and household utensils were still being produced from wrought iron, made in bloomery furnaces. Between the blast and bloomeries, the consumption of charcoal in England reached massive proportions. The effects can still be seen in one area where iron making flourished – the Weald of Kent.

The Weald today is a broad expanse of gently rolling hills and vales, a patchwork of fields and hedgerows, with scattered woods and copses. Some of the fields are under pasture, and carry herds of dairy cattle. Others grow the hops for which Kent is famous, and the spires of the village churches are echoed by the tall conical roofs of the oast-houses, where the hops are dried. The low profile of the Weald has a timeless, unchanging look. But this is deceptive. The landscape today is not the original countenance of this part of southern England.

Once, the Weald was densely forested with mature oaks, beech and chestnut. But the forests of the Weald stood on chalk, and in the chalk, beginning in Roman times, rich pockets of iron ore were found. In almost every surviving patch of woods there are shallow pits, where first the Romans, then the Saxons, and later the iron makers of medieval England dug out the iron. To smelt and forge the iron, however, the ironmasters needed charcoal — and so began the clearance of the forests.

By the sixteenth century the Weald was dotted with blast furnaces and forges, each surrounded by an expanding patch of cleared land. Most of the furnaces were built beside small streams, and power was provided by a waterwheel, driven by water stored in a small dam. A typical site lies at the foot of Furnace Lane in the village of Brenchley. The foundations of the wheel and its waterway can still be seen in the stream bed beside tree-lined banks. The dam, overhung with willows, now provides carp fishing for the villagers. The cannons cast here were so famous that they were exported to many countries across the Channel. There is an echo of that trade in the name of the Gun and Spit-Roast Inn in Brenchley.

However, as the iron industry grew it generated not only trade but conflict. The forests of Britain had long been admired for their most celebrated inhabitants, slow-growing and long-lived. These were the mighty oaks, which for centuries had provided the wooden walls of old England. The crucial importance of the sea defences of the British Isles was first acknowledged by Henry VIII, who in effect founded the Royal Navy, and built the first Royal Dockyard at Portsmouth. In that dry dock lies the most majestic survivor of a long line of wooden warships reaching back to Alfred the Great – HMS *Victory*, Nelson's flagship at the Battle of Trafalgar. She contains two thousand one hundred tonnes of oak. Thousands of trees – a whole forest – had to be cut for a ship of the line like *Victory*, and the same huge tonnages of timber were needed for the ships of the growing maritime trade.

So two uses of oak were in conflict. And now a third industry added its demands to the pressures on the forests: glass making. Such was the demand for window glass in the sixteenth century that scores of glass makers from Europe crossed the Channel to England and set themselves up in business in the forests, where they found timber for their furnaces. The government was finally forced to act.

In 1558 a law was passed forbidding 'the felling of trees to make coals for the burning of iron', but the Weald of Kent and Sussex was exempted, perhaps because of lobbying by the thriving iron industry in that area. And still the price of wood continued to climb. In 1559 a writer complained that the price had risen from a penny to two shillings a load 'by reason of the iron mills'. By 1581 the shortage of wood for ship building was so serious that a further act was passed, forbidding the felling of trees within twenty-two miles of the Thames, within four miles of the great forests of the Weald, and within three miles of the coastline, anywhere. This decree effectively wiped out the iron-making industries of the Weald, but the consumption of wood by the glass makers continued, and so did the domestic demand from city dwellers.

By 1615 England was facing an energy crisis. A Royal proclamation in that year lamented the disappearance of the kind of wood 'which is not only great and large in height and bulk, but hath also that toughness and heart, as it is not subject to rive or cleave, and thereby of excellent use for shipping'. And so, at last, that country was forced to turn to a source of fuel which had been known for centuries, but which few had chosen to use: coal.[43]

It is impossible to say when coal was first burned deliberately. It made little contribution to the technologies of the ancient world in the Near East or the Mediterranean, because there were no large outcrops of coal in those regions. The first recorded accounts of the regular use of coal as fuel come from China, during the Han dynasty (206 BC to AD 220). The Chinese seem to have been consistent users of coal for millennia, because Marco Polo, on his return to Italy from China in the thirteenth century, reported that in places the people burned 'black stones' for fuel.

The first known use of coal in Europe was in Britain, during the Roman occupation, when pieces were picked up on the sea shores, and particularly along the Firth of Forth in Scotland. Because of this association it was called 'sea coal', although we know now that it had been broken away by the waves from the seams which outcropped along the coast. Once the Romans left in the fifth century AD, however, coal seems to have been forgotten for centuries.

There were attempts to use coal for heating houses after the Norman conquest of Britain, but the sulphurous fumes and smoke were intolerable in ordinary houses, which at that time had no chimneys. The fireplace was simply a depression in the middle of the floor, and at a certain time each night, at the ringing of a bell, the fire had to be put out by being covered – the *couvre feu* in French, which became curfew. In succeeding centuries, however, as the price of wood, charcoal and peat rose, there was a renewed interest in coal, especially when houses began to include fireplaces and chimneys.

By the Middle Ages the British had become the world's largest users of coal. Rich seams had been discovered in many parts of England, Scotland and Wales. Many industries, heavy users of fuel, had converted to coal because of the cost of charcoal. For some, such as glass making, this had been made possible by a significant advance in technology: the development in 1612 of the 'reverberatory' furnace. In this, the materials to be heated were not mixed with the fuel, as previously, but placed in crucibles at one end of a tunnel-shaped furnace, near the chimney. The hot gases from the burning coal passed through the tunnel, and heat was reflected from the arched brick roof on to the crucibles. The materials they contained were therefore not contaminated by contact with the noxious elements in the coal.

Only one important industry in Britain had not been able to convert to coal, and that was iron making. The reverberatory furnace had been adapted for the smelting of copper, tin and lead, but iron ore could not be smelted in this way. As for the blast furnaces, attempts had been made to use iron ore mixed with coal instead of charcoal, but without success. The problem was not only the sulphur, phosphorus and other contaminants in the coal, which tended to make the iron brittle and useless, but the nature of the coal itself. It was softer than charcoal, and as it burned it 'slumped' into a dense mass which choked the furnaces. Iron makers were forced to go on using charcoal, regardless of its still rising price. The inevitable result, towards the end of the 1600s, was that Britain, an island rich in both iron ore and carbon fuel, found itself suffering industrial stagnation. The urbanisation of society and the growth of the economy were frustrated by the restriction of supplies of the key metal, iron.

There is no surviving portrait of the man who found the answer to this problem, and in doing so released the pent-up forces of the Industrial Revolution which swept England, then Europe, and finally the world. His name was Abraham Darby, and he was a Quaker with no particular training in metallurgy. In fact, like many long-delayed advances in technology, Darby's discovery involved nothing startlingly new. The separate elements of his innovation were well known, and would inevitably have been brought together in the same way by someone else, had he not done it. But the fact remains that Abraham Darby was the first man to smelt good quality iron with a fuel derived from coal rather than wood.

Abraham Darby had been apprenticed in his youth to a firm in Birmingham which made small brass mills for grinding malt, used in the brewing of beer. During those years he had become familiar not only with metal casting but with the malting process, including the method of drying the malt in ovens. The fuel used by the brewers was not charcoal, and nor was it coal. It was something devised to meet the requirements of malting – and its particular properties seem to have impressed themselves upon young Darby's memory.

In 1702 Darby went to Bristol as a partner in the Brass Wire Company. At this time brass – the alloy of copper and zinc – was becoming extremely popular in Europe for all kinds of articles. It could be cast, machined, and shaped with ease, and its bright appearance was appealing. In 1704 Darby visited Holland, hoping to recruit Dutch brass workers and their skills with water-driven trip-hammers. While there he saw the Dutch making large brass pots by casting them in sand moulds. He conceived the idea of making cooking pots from cast iron, which would be considerably cheaper.

On his return to Bristol, Abraham Darby made some experiments in the techniques of moulding and casting thin-walled pots in iron, and then looked around the West Country of England for a suitable location. He found it in a small town called Coalbrookdale, on the upper reaches of the River Severn, in Shropshire. Darby moved there

The brick furnace at Coalbrookdale in Shropshire that was used by Abraham Darby in 1709 to make the first iron ever smelted successfully with coke instead of charcoal. *Three generations of the Darby family eventually used this furnace, which now forms part of the Ironbridge Gorge Trust Museum in the valley of the Severn, near Shrewsbury.*

in 1707 and leased a disused blast furnace. He and his apprentice, named John Thomas, let it be known that they proposed to make iron cooking pots and, what is more, to make them without using charcoal, which by then was so expensive as to be quite uneconomic.

Darby himself never explained exactly why he chose Coalbrookdale, but the reason usually given is that he knew the local coal was low in sulphur, an element which had always been a problem in making iron with coal. What may also have influenced Darby, however, was the fact that Shropshire coal had physical properties which made it ideal for making into coke. This was the fuel which Darby remembered being used by the brewers of Birmingham for making malt. What worked for malt, Darby may have thought, might work for iron.

Coke was not widely known as a fuel at that time, chiefly because its use was more or less restricted to the brewing industry, where it had been developed to meet a particular problem. Like other early industrialists, the brewers had switched from charcoal to coal for their furnaces at the beginning of the seventeenth century, but it had proved unsuitable for malting. The sulphur in the coal tainted the malt, and the customers refused to drink the beer made from it. Fortunately, an entrepreneur named Sir Hugh Platt had suggested, in 1603, the idea of charring coal to rid it of its impurities, precisely as wood was charred to make charcoal. Platt's plan was to make briquettes to 'sweeten' the smoky domestic fires, but at first no successful method of charring coal could be found. It was not until about 1642 that the brewers of Derbyshire succeeded in partly burning coal in covered heaps. The resulting 'cinders' were harder than coal, consisted of almost pure carbon, and burned with a clean, smokeless heat. They were called 'coaks', and malt dried with 'coaks' or coke made sweet, pure beer, which became famous throughout England.

However, the possibility of using coke for smelting iron did not apparently occur to anyone other than Abraham Darby. He seems to have worked on the supposition that the same volatile constituents of raw coal which made beer unpalatable might also be making cast iron brittle and useless. It is not known whether Darby also realised that the structural rigidity of coke would be an advantage in keeping iron furnaces from clogging during smelting.

And so it happened that Abraham Darby prepared some coke from Shropshire coal, mixed it with local iron ore, and charged his furnace at Coalbrookdale. On 4 January 1709, he and his assistant blew the furnace with bellows for some hours, and then Darby tapped it. Out through the brick arch of the furnace flowed the first iron of quality ever made from coal.

News of Darby's success was slow to spread, but when it did it caused intense interest throughout England. Coalbrookdale became the

Coalbrookdale became the centre of the iron-casting industry in England and produced a wide range of articles. *This stove is in the display at the Coalbrookdale Museum of Iron, beside Darby's furnace.*

centre of the British iron industry, as new furnaces and foundries were set up there. One of the first is still casting iron stoves in sand moulds, just as it did at the beginning of the eighteenth century.

Darby's achievement was enormous in its portent. He had shown that iron could be made with the apparently limitless resources of Britain's coalfields, rather than the depleted reserves of her forests. That low brick archway in Abraham Darby's furnace (which still stands on its original site, in what is now the town of Telford) was the gateway to a new Iron Age. He had created a new dynasty, whose fiery domain would extend the length and breadth of Britain: the ironmasters.

Iron stoves are still being cast by hand in sand moulds, exactly as they were in Darby's time, at the ironworks which has survived in Coalbrook-dale. *The recent revival of interest in fuel-efficient stoves has given this company a new lease of life.*

The widespread change to smelting iron with coke not only freed Britain from industrial stagnation, but immediately multiplied the demand for coal. By this time, however, the coal mines themselves were facing an obstacle to production which had grown progressively worse over the centuries. As the surface seams had been dug out, and the mines compelled to go deeper, they had come up against the miner's ancient and most formidable enemy: water.

As already described, the problem of water seeping into the mines from underground water tables or aquifers had been recognised as early as Greek and Roman times. But despite the remarkable technology of Roman engineers such as Vitruvius they had been unable to provide a permanent answer. With the resurgence of mining in Europe after the Dark Ages, all kinds of pumping systems had been tried, driven by every known source of energy, from windmills to human slaves. None had proved adequate, for they all suffered from one crippling, insurmountable difficulty. They lacked a reliable source of continuous power.

154

It is difficult for us to grasp now, but as recently as AD 1700 there was nowhere in the world any source of power for the continuous driving of machinery other than wind, water or living muscle. By the sixteenth and seventeeth centuries many powerful minds were working on this challenge, and the physical principles upon which success was eventually won had become known. But the combination of the necessary strands of research into a workable power system had been delayed by a particular attitude of mind in Europe at that time.

By founding the Royal Society in 1660 Charles II had conferred the seal of royal patronage and approval upon the growing spirit of scientific enquiry which, in England as in Europe, typified the Renaissance. But during the seventeenth century this ferment of the imagination was still confined almost exclusively to the study and the laboratory bench. Knowledge was pursued for its own sake, and although very significant discoveries were made, the discoverers were seldom concerned to apply their knowledge to practical purposes. When the seventeenth century scientists needed instruments or apparatus, they turned to the great clock makers of the day. Throughout that century the beautiful work which these master craftsmen executed in this new sphere remained the only practical evidence that a new spirit was abroad. The consequences of this division have been described by L. T. C. Rolt and J. S. Allen:

> Between this small and exclusive band of craftsmen and the makers of all those practical things by which man seeks to lighten his daily toil an immense gulf was fixed. For example, the makers of wind and watermills built solidly and well, but with such massive crudity and ignorance of principle that it is almost impossible for us to believe that workmanship so medieval in character could be contemporary with that of the great horologists and instrument makers of seventeenth century London. For many years there was no bridging this gulf. Instrument making was a 'mystery', a jealously guarded closed shop, while the scientists of the day looked upon the practical man, who lived by the skill of hand and eye, with a certain arrogance and contempt that is typified in their attitude towards Thomas Newcomen.[44]

Newcomen's name has been overshadowed by other, more illustrious names in the history of the Industrial Revolution. His grave, which lies somewhere in London, has been lost from record. Today, however, there is growing recognition of his enormous contribution to that radical transformation of human existence. For Thomas Newcomen was responsible for perhaps the single most important invention in the history of technology: the first power-driven, self-regulating machine, which would keep working as long as it was supplied with fuel.

The details of Newcomen's life are not well known, for before his great achievement – and even after it – he was not much noticed, least of all by the scientific 'establishment' in London. From what we do know, it appears that Thomas Newcomen was born in 1663, the son of a merchant in Dartmouth, a small fishing port in Devon. When he was about twenty-two he set himself up in business as an ironmonger in Dartmouth. In those times, a tradesman of that description did not just sell tools, but performed many of the jobs of a blacksmith and a craftsman. It was in this capacity that, in the early 1700s, Newcomen made frequent visits to the tin and copper mines of Devon and Cornwall. Thus he must have become aware of the problem of water in the mines, and of the great need for some kind of mechanical device to pump them out.

What persuaded Newcomen to turn his mind to the potential of steam as a source of energy in this matter remains unknown. Nor is it clear how much he knew of the experiments which had been carried out in various parts of Europe with the newly discovered but little understood force of atmospheric pressure. Newcomen was, however, a curious as well as practical man, and was aware of such publications as the *Philosophical Transactions*. Dartmouth was an important seaport, and had good communications with the rest of Britain and the Continent. It must therefore be assumed that Newcomen drew upon the existing 'state of the art' in building his engine.

Three lines of development would have been of particular interest to Newcomen. One was represented by the early attempts to harness the power of steam, notably by Gianbattista della Porta in Naples. In 1606 della Porta published a drawing showing water being forced up a pipe by steam from a closed boiler. In 1682 Charles II in England was greatly impressed by an invention, based on the same idea, for raising 'any quantity of water to any height by the help of fire alone'.

In 1672, meanwhile, Otto von Guericke in Germany had performed two historic experiments, involving the then mysterious interaction between a vacuum and the pressure of the atmosphere. In the first, he showed that when two copper hemispheres were pressed together and the air evacuated from them, they could not be pulled apart by sixteen horses. In the second he sucked air from a cylinder containing a piston, and twenty men holding a rope passing over a pulley could not prevent the piston from being forced to the bottom of the cylinder. Thus von Guericke convincingly demonstrated that the atmosphere exerts a strong pressure.

Some time around 1695, a French Huguenot refugee in England, Denis Papin, worked out how a piston in a cylinder could be made to perform work by a combination of atmospheric pressure and steam. When Papin filled the cylinder with steam, and then allowed it to condense, the atmospheric pressure on top of the piston forced it down, thus raising a weight attached to a rope passing over a pulley.

The catalyst which eventually prompted Newcomen to synthesize all these stumbles into one huge step forward may have been his association with Thomas Savery, another Devon man, and a prolific inventor. Savery actually built and patented, in 1698, a device which used steam in a boiler to force water up a pipe. The Savery 'pump' was tried out in Cornwall, but as it had to be located at the bottom of the shaft, and had a limited lift of about fifteen metres, it was unsuitable for pumping out deep mines. It also needed steam under high pressure, and the technology of boilers and pipes at that time was not capable of handling such pressures safely. Savery lost interest in his engine about 1705, but different versions of it were tried for a long time afterwards. As far as

Opposite: One of the most significant advances in the history of technology was Thomas Newcomen's 'atmospheric engine' – the world's first source of continuous power other than wind, water or living muscle. *This is a contemporary drawing of Newcomen's first recorded engine, which was installed at a colliery near Dudley in Staffordshire in 1712.*

156

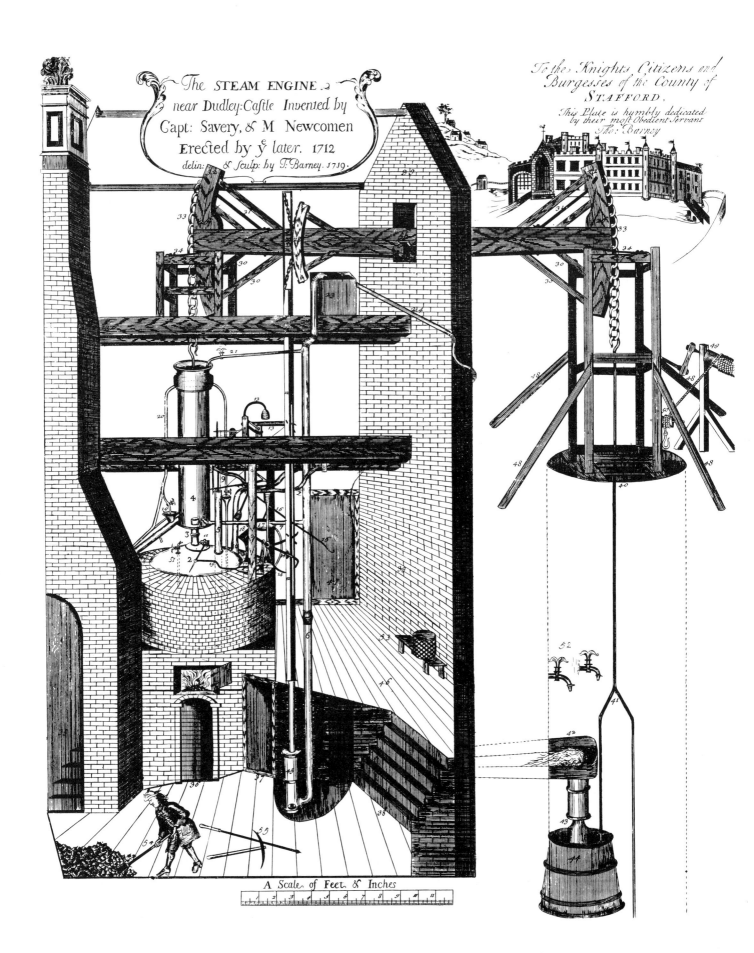

The STEAM ENGINE
near Dudley:Castle Invented by
Capt: Savery, & M.ʳ Newcomen
Erected by y.ᵉ later. 1712
delin: & sculp: by T. Barney. 1719.

To the Knights, Citizens and
Burgesses of the County of
STAFFORD.
This Plate is humbly dedicated
by their most Obedient Servant
Tho: Barney

A Scale of Feet & Inches

the central problem of mine pumping was concerned, however, a comprehensive solution was close at hand.

It came from Thomas Newcomen and, like Gutenberg's printing system, it seems, remarkably, to have emerged fully developed. If there were any experimental versions, or preliminary scale models, they have not survived. Nor is there anything other than circumstantial evidence of the lengthy development work that Newcomen must surely have put into his device in the mines of Cornwall. The first known Newcomen engine began pumping water out of a coal mine near Dudley, in Staffordshire, in 1712. It was of impressive dimensions, and pumped 'one hundred and twenty English gallons' of water a minute from the bottom of the mine shaft, which was fifty metres deep.

'The fame of this excellent pumping engine soon spread across England,' wrote a contemporary author, Marten Triewald, 'and many people came to see it, both from England and from foreign nations. All of them wanted to make use of the invention at their own mines, and exerted themselves to acquire the knowledge needed to make and erect such a wonderful engine . . .' From the detailed drawings of the engine, published in 1719, it is obvious how Newcomen had succeeded where all before him had failed in making a workable engine. What he had done was to combine von Guericke's and Papin's idea of the piston and cylinder with Savery's steam boiler.

In Newcomen's engine a cylinder, open at the top, was mounted vertically above a boiler. A piston inside the cylinder was connected by a chain to one end of an overhead beam, pivoted in the centre. From the other end of the beam a rod descended into the mine shaft, where it was attached to a pump. At the beginning of the working cycle the weight of the pumping gear pulled down its end of the beam, and raised the other end. This brought the piston to the top of the cylinder, in the process sucking steam into the cylinder from the boiler. A jet of cold water was then injected into the cylinder. This caused the steam to condense instantly, producing a partial vacuum under the piston. The atmospheric pressure on top of the piston immediately forced it down. This action pulled down the beam, and operated the pump attached to the other end. The cycle was repeated every five seconds, for as long as the boiler produced steam.

It will be appreciated immediately that what Thomas Newcomen had invented was not, as is sometimes suggested, the first 'steam engine'. Its power came not from the pressure of steam, but from the pressure of the atmosphere, working on the partial vacuum produced by the condensation of steam. But Newcomen's 'atmospheric engine' was man's first mechanical source of continuous energy, the first replacement for animal muscle that was independent of the elements – and it was devised by a man without any technical training.

Newcomen was not a scientist, not an engineer, but that archtype of the Industrial Revolution, the 'practical man', with the insight of genius. T. E. Lawrence – Lawrence of Arabia – wrote of genius that when it happened 'it was like the flash of the kingfisher across the darkened pool, vivid and unmistakable'. That is how it was with Thomas

Newcomen's solution to an age-old problem, which had formed an insuperable barrier to the wealth within the earth.

As an example of mechanical design, Newcomen's engine was crude and inefficient. Less than onc per cent of the heat energy used in raising the steam was converted into mechanical energy. This meant that the engine used a great deal of fuel, such as coal. But the need for a mine pump was so desperate that it over-rode immediate questions of fuel cost, and in any case coal was extremely cheap at the mines.

Apart from Newcomen's insight into the relationship between steam and atmospheric pressure, one reason why he succeeded where others had failed was that he did not ask the impossible of contemporary technology. The steam pressure in his cylinder, as it was drawn from the boiler, was barely above atmospheric pressure. The piston did not need to fit the cylinder very closely, and was sealed very simply with a layer of water on top of it. The beam transmitted great power, but slowly. Thus there was no necessity for sophisticated bearings.

The simplicity and directness of Newcomen's concept can still be appreciated in his home town of Dartmouth, where one of his very earliest engines is preserved in working order. This engine was built around 1720 for pumping out a coal mine in the Midlands, which it did for just on a hundred years. In 1821 it was installed on a canal near Coventry to pump water out of a well into the canal. It worked at this task for almost another century, until 1913, when it fell into disuse. In 1963 it was retrieved by the Newcomen Society and taken home to Dartmouth.

Left: A diagram of the simple but effective manner in which Newcomen, having created a partial vacuum beneath the piston by condensing the steam in it with a jet of cold water, used the pressure of the atmosphere to drive down the piston of his engine. *Right*: An early Newcomen engine, built in 1760, which was abandoned earlier this century and lay derelict until 1930, when it was acquired by Henry Ford for his museum at Dearborn, near Detroit. *Both illustrations are from* The Steam Engine of Thomas Newcomen *by L. T. C. Rolt and John S. Allen (see Notes page 266).*

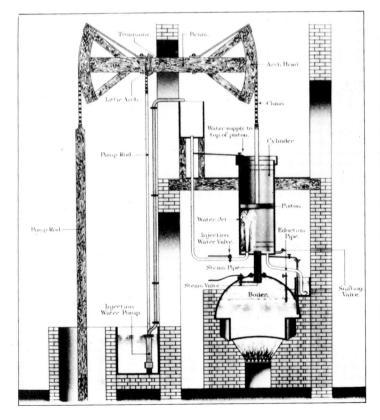

Today the engine stands in a small building in the municipal park, near the harbour. Its iron boiler was broken up for scrap during World War II, and the fifty-five centimetre diameter piston – on which the atmosphere exerted a force of about two tonnes – is now operated by hydraulic pressure. Otherwise it is just as Newcomen made it. The great wooden beam still rises and falls with its original majestic tempo of twelve strokes per minute, and it pumps water from a well beneath the building. Here is the oldest engine left in the world, the articulated skeleton of a tremendous achievement. With it, Newcomen not only found a way to pump out the mines, but ushered all mankind into the modern era. What he had created was the embryo of the powerful, tireless mechanical slave which would power the Industrial Revolution – the true steam engine.

Within a decade, the nodding heads of Newcomen's obedient work-horses could be seen drinking deeply at mines all over Britain, and even across Europe. The Tyne coal basin in the north of England was in particular need of pumping engines, because of the depth of the coal seam, and bought them eagerly, despite Newcomen's high patent fees. In Cornwall, where Newcomen had first begun to apply his mind to the problem of drainage, his rocking beam was soon to be seen everywhere. By the time Newcomen died, in 1729, nearly one hundred of his engines were in use.

Newcomen's original patent ran out in 1733, and engineering firms everywhere began to build their own versions. In Cornwall, in particular, the combination of pressing needs and growing experience produced ever larger engines. In its final manifestation, the mighty Cornish pump was one of the mechanical wonders of the world. The largest had cylinders and pistons more than two metres in diameter, and massive cast iron beams weighing up to fifty tonnes. The pump rods in some mines extended four hundred metres down the shafts, and the weight of the pumping assembly reached one hundred tonnes. To house the engines and take the immense strain of the pumping operation the Cornishmen evolved their tall and characteristic stone engine houses, which in the century which followed Newcomen's invention were to become known all over the world. In Cornwall itself more than fifteen hundred engines were built for the mines. The last one working shut down only in 1955.

The strength and reliability of the Cornish beam engine led to its being adapted to the task of pumping water in situations far removed from mining. The most spectacular examples of these giants still to be seen in operation are those in the Kew Bridge Engines Trust and Water Supply Museum in London. The five huge beam pumps, with cylinders ranging up to nearly three metres in diameter, were installed from 1837 onwards to pump water from the Thames into London's mains supply system. The two largest engines, built in Cornwall, had a piston force of more than a hundred tonnes. When they were installed at Kew in 1869 they represented possibly the greatest concentration of mechanical power in the world. Between them they lifted eighty-two thousand tonnes of water a day.

These huge monsters worked tirelessly for nearly a century, until

Opposite: This working model of a Newcomen engine was given to the instrument maker at Glasgow University in 1763 for repair, with momentous consequences. That instrument maker's name was James Watt. *The model is now in the Glasgow University Museum.*

160

1944. They have been restored by the Trust, and when they are steamed and worked at weekends they form a suitably impressive tribute to the memory of Thomas Newcomen. In action they are a majestic sight, but remarkably quiet. The hushed atmosphere of the lofty engine house, as tall as a cathedral, is broken only by the clank of the self-regulating control rods, the distant rush of water through the pump barrels, and the heavy breathing of these dinosaurs of the Industrial Revolution.

In tracing the evolution of the beam pump to its final imperial dimensions we have, in a sense, been following just one isolated branch of the family tree of steam power which grew out of Thomas Newcomen's first crude device. Yet such was the magnitude of his achievement, and the simple practicality of his design, that for fifty years it was susceptible to no improvement. In fact it was not until 1763, when the instrument maker at Glasgow University was given a working model of a Newcomen engine to repair, that the basic concept of the 'atmospheric engine' was questioned. That instrument maker's name was James Watt.

Watt is widely regarded as the 'inventor' of the steam engine, following his supposed observation of the lid of a boiling kettle being lifted by steam pressure. In fact, it was the Newcomen model engine which introduced Watt to the potential of steam. The improvements he made on Newcomen's design, however, were the products of his own great insight, and they brought into existence the new source of energy which, in all its forms, was to transform the nature of industrial production, transport and travel, and in doing so release a great part of mankind from the burden of physical labour.

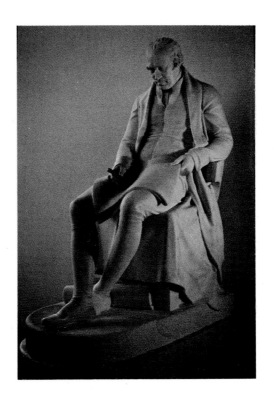

James Watt's analysis of the shortcomings of Newcomen's 'atmospheric engine' led to his design for the first true steam engine. *Statue of Watt at Glasgow University.*

162

What struck James Watt immediately he came to study the Newcomen model engine was the great waste of heat energy in the working cycle. After the injection of cold water had condensed the steam in the cylinder and pulled down the piston, much of the heat of the incoming steam was used in warming up the cylinder walls again before the next stroke. In fact the model would stop working after a few strokes, because the boiler kept running out of steam. Since thermal losses in an engine increase with any decrease in size, the model was exaggerating the inefficiency of Newcomen's design.

Watt perceived that it was heat loss alone that was preventing the model from working. He applied his mind to the problem, using the new theory of the latent heat of steam worked out by his friend Dr Joseph Black, a professor of chemistry at Glasgow University at the time. First, Watt tried keeping the cylinder from cooling down by enclosing it in a hot water jacket, but this was not enough. His eventual and most significant modification was to provide a separate vessel, opening off the cylinder, in which steam could be condensed with cold water. The partial vacuum in the condenser drew the remaining steam out of the cylinder, until it had all been condensed. Finally, to prevent heat escaping out of the top of the cylinder he closed it in. Since the atmosphere could no longer exert pressure on the piston, Watt diverted steam at about the same pressure to do this work.

Watt went into a successful partnership with a shrewd Birmingham businessman named Matthew Boulton. Soon Boulton and Watt steam engines dominated the industrial scene, by demonstrating their great advantage in fuel consumption over Newcomen's original design. As Boulton told George III in a famous remark: 'I sell, sire, what all the world desires – power.' So efficient were the Boulton and Watt engines that some are still working. In fact one of them, pumping water on the Kennet and Avon Canal in Somerset, is the oldest steam engine in the world still performing the job it was installed to do. It began working in 1812 – the year that Napoleon marched on Moscow.

While the steam engine was a critical trigger factor in the surge of industrial expansion which began in the eighteenth century in Britain, of much wider significance was the chain reaction that it set crackling through the whole arena of the Industrial Revolution, by its stimulus of such vital activities as the production of coal and iron, the evolution of machine tools, and the development of new forms of transport for both goods and people.

The initial success of Newcomen's engines in draining the mines had produced an immediate increase in coal production, and the inevitable demands from the mine owners for more powerful pumps, to enable the shafts to go even deeper. The only way the engine builders could do this was by enlarging the cylinders – but here Newcomen and the other Cornish engineers ran into a serious problem. The first engines had used cylinders cast in brass, and since there were no suitable boring machines their inner surfaces had to be laboriously finished by hand. The cost of making very large brass cylinders in this way would have been prohibitive.

Left: The Cornish beam engine, based on Newcomen's original concept, reached its final dimensions in the massive pumps at Kew, which pumped water from the Thames into London's water supply for more than a hundred years until the 1950s. *Above*: There is no surviving portrait of Thomas Newcomen, but among his most enduring monuments are the tall stone engine houses over the abandoned copper and tin mines of Cornwall.

Fortunately, Newcomen had reason to visit Coalbrookdale, and discovered that the iron-casting industry developing there, following Abraham Darby's introduction of coke smelting, was quite capable of casting cylinders in iron of almost any size he ordered, and at a price that was highly competitive with brass. And so the pumping engines grew in size, with their cast-iron cylinders and pistons, until the Coalbrookdale foundry was casting cylinders more than three metres long and nearly as great in diameter.

With the low steam and atmospheric pressures involved in the Cornish beam engines, the relatively poor fit between the cast-iron cylinders and their pistons was unimportant. Once James Watt developed his steam engine, however, this matter of tolerances became critical. When Watt obtained his patent in 1769 there was no method of making a tight-fitting piston and cylinder, other than casting them as accurately as possible, and finishing by hand. This was a tremendous obstacle to the wider adoption of the steam engine until, as so often happened in those days of practical engineering, the need produced the solution.

In this case, impetus came from an unexpected quarter – the French army, where, it was said, the soldiers 'feared their own cannon more than those of the enemy'. The reason, apparently, was that most cannons at that time were cast in one piece to their finished size and bore. Since hollow castings were not particularly strong, they frequently blew up, taking their gunners with them. A French army delegation to England mentioned their problem to the ironmasters. One of them, John Wilkinson (who was such an enthusiast for all things iron that he wanted to be buried in an iron coffin) set out to make a stronger cannon from cast iron.

Wilkinson finally succeeded in 1774, patenting a method of cannon making which was entirely new. He first cast a solid cylinder of iron and mounted it on the axis of a wheel. This was turned slowly by a steam engine, while a hole was bored down the centre of it with a tool mounted on a well-supported shaft, and moved steadily forward by gears. The technique was soon applied to the boring of cylinders for steam engines, with an accuracy measured in fractions of a millimetre. In an age when most things were still fashioned from wood, the introduction of automatically guided precision tools was a revelation. In effect, it laid the foundations of the whole technology of manufacture by machine tools, upon which the modern world is built.

The improved Boulton and Watt steam engines, with their precision-bored cast-iron cylinders, now returned the benefits of increasing scale and efficiency to the iron industry itself. Since the fifteenth century the ironmasters had relied upon waterwheels to pump the bellows to provide the blast for their furnaces. But as more furnaces came into blast to meet the growing demand for iron, the supply of water available for turning wheels became totally inadequate. The first reaction was to use the water over and over again. Horse-driven pumps and then even Newcomen engines were used to return water from below the waterwheels to the reservoirs above them.

Eventually, the leather bellows which had blown the blast furnaces were replaced by huge iron piston-driven air pumps. When these were finally coupled to large Boulton and Watt steam engines, the smelting of iron was revolutionised. No longer dependent upon water resources, the blast furnaces multiplied, and the roar of their blowing engines echoed across the countryside day and night.

CHAPTER 8
Steam and steel launch the revolution

By the last quarter of the eighteenth century, the many components of the mighty engine of the Industrial Revolution were working together to generate an irresistible surge of production and economic growth in the British Isles.

Everywhere, perishable contraptions of wood and leather were being replaced by enduring machines of iron and brass. The natural forces of wind and water and muscle had been largely superseded by steam, tireless and obedient. The conversion of the back and forth action of the piston into rotary motion by the introduction of the crank multiplied the application of steam throughout industry. James Watt's invention of the centrifugal governor, based on a device used to regulate the speed of grindstones in wind-driven flour mills, finally provided a degree of control over engines that was foolproof and unfailing. As the first automatic feedback mechanism, Watt's governor must be regarded as the distant herald of automation.

Among the first of the manual enterprises to be seduced by the new technology was the ancient craft of spinning and weaving. This cottage industry was already being forced into the mills by water power and had been mechanised to some degree by the invention of the spinning jenny by James Hargreaves and the loom and carding machine by Richard Arkwright. The rotative steam engine gave the final impetus to the growth of the textile belt across Lancashire and Yorkshire, and thousands of such engines were soon turning spindles and driving flying shuttles. Textiles became Britain's greatest export earner, and it is from the mechanisation of this industry that many historians date the beginning of the Industrial Revolution.

Many steam engines in the cotton and woollen mills went on to perform their tasks for a century or more without breakdown or repair. Some survived long past the time when innovations elsewhere had overtaken the textile industry itself, and a few were still working in the early 1980s.

Above: One of the first industries to be mechanised during the Industrial Revolution was spinning and weaving. *This cotton mill in Lancashire was still worked by steam until it finally closed down in 1981. Left*: The ruins of the Bedlam furnaces on the bank of the Severn in Shropshire, which once turned the night sky of Coalbrookdale red. *These are the furnaces whose ruddy glare lights up the dust jacket and frontispiece of this book.*

No less significant than the impact of steam on the textile industry was the stimulus that it gave to coal mining in the eighteenth century. After beam pumps to drain the shafts came steam winding engines to lower the men and raise the coal. And as the shafts drove deeper into the rich black seams lying under the British Isles, supplies of coal for iron making became virtually unlimited.

In Coalbrookdale itself the iron foundries worked around the clock. It was an amazing scene of activity. The red furnace glare against the night sky, the stupendous roar of the blasting engines – compared by observers to the thunder of massed guns – were an assault on the eyes, ears and mind. In the words of one visitor, 'the smoke and fumes and glare of the furnaces named Bedlam were like a descent into Hell itself. It needed only Cerberus to make it look like a heathen Hell!'

Caught up in the momentum of the Industrial Revolution, the ironmasters of Coalbrookdale were full of exuberance and the spirit of adventure. There seemed no limit to what their industry and their iron could do. So when, in 1777, a group of local citizens proposed an undertaking that no other iron-casting centre in Britain could contemplate, and which would therefore spread the name of Coalbrookdale afar, it was seized upon. Its execution was entrusted to Abraham Darby III, the grandson of the Darby who had first made iron with coke.

Abraham Darby III was now managing the Coalbrookdale Company, which had grown out of his grandfather's business. He quoted a price of three thousand two hundred pounds for the job, and when this was accepted he ordered his grandfather's original brick furnace to be enlarged. On open ground beside the furnace a special earth floor was laid down, and on it Darby's men prepared open sand moulds larger than any ever used in Britain before. When all was ready, and before expectant crowds, the furnace was charged and blown, and Darby gave the order to pour. The iron which flowed into the long, narrow, curved sand troughs was the first of a total of nearly four hundred tonnes of metal which were cast over the next six months. It went to create the ribs and struts of what was to become the first undoubted masterpiece of the Industrial Revolution – the Iron Bridge.

The Iron Bridge was not only the first bridge in the world made of cast iron; it was the first large work of civil engineering in iron. The audacity of the men of Coalbrookdale was undeniable. When they decided to build an iron bridge there was no thought of trying out the technology with a cautious foot-bridge over a canal. With magnificent boldness, and in defiance of all contemporary theory and warnings, they threw their bridge across the Severn in Coalbrookdale, in a single graceful span of one hundred feet (about thirty metres). The arch took only three months to assemble from its pre-cast sections, which included five massive ribs, each twenty-one metres long. Stone abutments and approach roads were added, and the bridge was opened to the public on New Year's Day, 1781.

Among the thousands who came to admire the Iron Bridge was John Wesley. He had preached a sermon in Shrewsbury the previous evening, and then set out to walk along the towpath of the Severn to see 'the first brilliant audacity of the ironmasters'. Wesley was greatly

impressed, he confided later to his diary, and with nothing else in the English landscape to measure it against, he saw it in the classical perspectives of the ancient world and compared it to the Colossus of Rhodes.

The Iron Bridge still spans the Severn as gracefully as ever, the progenitor of all the many forms of metal bridges that followed it. It expresses not only the innovative thinking but the tradititional Shropshire caution of those 'practical men'. At that time, the use of iron in structures of this kind was completely untried, so the architect and the builders fell back, for reassurance, on the proven techniques of wood working. There is not a bolt or rivet in the main arch. The ribs are located in the iron base-plates with dove-tailed mortice joints, while the struts pass through slots and are chocked in place with iron wedges.

The Iron Bridge is part of an evocative landscape of that period of enterprise and adventure when Coalbrookdale was the iron-making centre of the world. Along the banks of the Severn still stand massive monuments to what Lewis Mumford called the profound shift 'from the technology of wood, wind and water to the technology of carboniferous capitalism'. They include Abraham Darby's furnace, in the grounds of the only ironworks still working; the huge ruins of the Bedlam furnaces; the complex of foundries and tall stone buildings on Blists Hill

The Iron Bridge across the Severn near Coalbrookdale was the first undoubted masterpiece of civil engineering during the Industrial Revolution. Its builders used many of the techniques of the carpenter to link their cast iron sections, including dovetails and mortice joints, but the many graceful details showed a grasp of the potential of this new construction material that transcended sheer strength and rigidity. *The Iron Bridge still stands, as sound today as when it was opened to the public on New Year's Day, 1781.*

169

housing the great steam-driven blowing engines; the Coalport china-works; and the extraordinary ascent of the Hay inclined plane.

This latter installation was built in 1792 near the village of Hay to overcome the topographical problems which faced the ironmasters of Coalbrookdale. The first five ironworks to be built were clustered along a brook where it joined the Severn, so that the water could be used to turn waterwheels to blow the furnaces. However, the coal needed to make the coke came from pits in the hills to the north of the Severn. Canals had been built to bring the coal to the river, but by the time they reached Coalbrookdale there was a difference in level of more than sixty metres. It would have required at least twenty-seven locks to move loaded barges up or down, and the operation might have taken three hours. The inclined plane was the ingenious answer.

It was simply a long, paved slope from the river level to the upper canal carrying parallel pairs of iron tracks. The tracks entered the canal at the top and a sheltered pool leading to the Severn at the bottom. Barges were floated on to wheeled cradles running on the tracks and hauled up or lowered down the slope with the help of a steam engine at the top. Both sets of cradles were joined by ropes around a wheel at the summit of the incline, and the descending heavily-laden barges helped to haul up the empties. It needed only four men to exchange a pair of five-tonne boats in four minutes. (The principle of the inclined plane is still in use on the Rhineland canal system, where tanker barges of several thousand tonnes are handled.)

At the height of Coalbrookdale's activity there were five inclined planes in use in the area. They and the canals they served had been developed to cope with one of the major constraints on the swelling tide of productivity in the last quarter of the eighteenth century – the lack of an adequate transport system in Britain.

The parish roads of those times were little more than rutted tracks, almost impassable in winter and rough and dusty in summer. Most people had to walk to where they wanted to go, and the only goods transport between villages and towns was by pack-horses, wagon or coach. Travel was excruciatingly slow. It might take a coach or wagon several days to cover fifty kilometres. Costs were accordingly high. Coal at the Birmingham coalfields was eight shillings a ton, but transporting it only thirty kilometres doubled its price.

With the growth of the Industrial Revolution and the drift of population into the spreading cities, the transport problems became acute. There had to be a better way of moving the food, coal, iron ore, and other raw materials needed by the new industrial centres, and of distributing the increasing flood of goods and products to the waiting markets. It was this pressure that produced, over a short but hectic period of about sixty years, the extraordinary explosion of canal-building in Britain.

Water transport was one of man's earliest means of travel, and in the ancient world the development of artificial waterways was widespread and in some cases remarkably advanced. For example, the Suez Canal constructed by de Lesseps in 1869 had at least two predecessors linking the Mediterranean and the Red Sea – one built by the Pharaohs of Egypt in the second millennium BC, and another built by the Persian king Darius around 500 BC.

The natural river systems of Britain were first modified by the Romans. The Caerdyke and the Fossdyke were ditches which completed a navigable waterway from the River Trent through Lincoln to the sea near Boston. Until the Middle Ages, however, there was little other development, except for the straightening or clearing of existing waterways. The use of locks for moving a boat from one level to another was known, and in the seventeenth century new possibilities were opened up when news reached England of Leonardo da Vinci's concept of mitre-angled lock gates, which are kept shut by the water pressure. However, the rather hilly landscape of Britain and the differences in levels between rivers kept interest in canals to a minimum – until the Industrial Revolution brought demands for bulk transport and, in particular, the bulk transport of coal.

Coal was the first commodity that needed to be moved in large amounts and at low cost. The existing methods were inadequate, to say the least. An interesting comparison was made at the time to see how much weight could be moved by one horse-power, using various means of conveyance. A single pack-horse could carry no more than one-eighth of a tonne. Putting the load on wheels and tracks enabled the horse to move several tonnes. But load the cargo into a barge, and the horse could tow fifty tonnes with ease.

The first true canals were built to link otherwise isolated agricultural or industrial communities with navigable waters. A short channel from the St Helens coalfield in Lancashire to the Mersey, called the Sankey Brook Navigation, was the first of this kind in England. It was paid for by the merchants and town corporation of Liverpool, and was opened in 1757.

The canal system in England came into existence in the second half of the 18th century to move coal to the proliferating factories, to distribute the expanding production of manufactured goods, and to bring food to the growing populations of the cities. *Flights of locks, like this one near Birmingham, were necessary to cope with the hilly terrain across much of England.*

It was the Duke of Bridgewater who first became impatient enough with the transport situation to take decisive action. He owned coal mines at Worsley in Lancashire, but had great difficulty in getting the coal moved by river to his customers in Manchester. So he commissioned an engineer, James Brindley, to build a canal from his mine right into the centre of Manchester.

Brindley did what few had been willing to do before – he struck out across country, away from existing rivers. To avoid building locks he followed a contour line, even though this produced a rather round-about route. With great daring he carried his canal over the River Irwell on a stone viaduct supported by three arches, high enough to allow sailing barges to pass underneath. 'Vessel o'er vessels, water under water, Bridgewater triumphs – art has conquered nature!' exclaimed a commentator of the day.[45] When the Bridgewater Canal was opened in 1761 the price of coal in Manchester immediately fell by half. The canal revolution had begun.

Brindley next tackled a major project – a canal to link the Mersey, serving Manchester, Liverpool and the cotton towns of Lancashire, with the Trent and the woollen counties of Nottingham and Yorkshire. The Grand Trunk Canal, as it came to be called, started a frantic rush to build canals between all the major industrial centres of England, including London.

A new and innovative engineer, Thomas Telford, shortened the routes decisively by ignoring contours and driving straight across country, digging cuttings through hills and using the spoil to make embankments across valleys. Telford's dramatic cast-iron viaduct which carries the Ellesmere Canal across the Dee Valley at Pont Cysyllte, in nineteen spans more than forty metres above the river, remains one of the great engineering achievements of all time.

By the 1820s Britain was criss-crossed by an amazing web of canals, totalling nearly five thousand kilometres. They were alive with barges distributing the still growing traffic in coal, raw materials, food and manufactured goods. Unfortunately, this remarkable network of water-ways, the product of immense skill, effort, imagination and investment, did not form a system. The canals were built at many different contour levels, and although attempts had been made to link them with flights of locks, some were cut off. A more serious handicap to cross-country transport was that the width and depth of the canals and locks varied absurdly.

The problem had begun with Thomas Brindley, the 'father' of English canals. He had made the Bridgewater Canal a comfortable four and a half metres wide, even though the coal barges which carried the coal from Worsley to Manchester were extraordinarily narrow – fourteen metres long but only one and a half metres wide. These 'starvationer' boats, as they were known, were used to bring the coal out of the Worsley mine by an astonishing system of underground waterways which extended through tunnels almost to the coal face.

Brindley then made the Grand Trunk Canal the same width, except for the centre section, which had locks only just over two metres wide. The reason for this strange decision was, Brindley said, economy in building locks, cutting tunnels and using water. Whatever Brindley's thinking, the long-term effect was disastrous.

With Brindley as consulting engineer, the network of canals which grew out from the Mersey and Trent to the West Country and the Severn, to Birmingham and Coventry, and to the Thames and London, all had the same narrow locks. So while other engineers were building wider canals and locks, which could handle barges carrying up to fifty tonnes, there was a band of canals across the centre of England which could take only narrowboats, carrying perhaps thirty tonnes.

The government of the time made a fatal mistake in not requiring the main trunk canals, at least, to be of practicable and uniform size. As it was, the narrow locks could hardly be other than crippling bottle-necks in the arterial circulation of trade and commerce, often necessitating slow and costly trans-shipment of goods from large barges to the picturesque but less capacious narrowboats. Added to this was an inescapable handicap of canal transport – it was slow.

Cheaper and more reliable than road transport it may have been, but canal transport was no faster. Lacking any other form of propulsion, it was tied to the pace of the plodding horse. And even this measured progress was intermittent, because of the frequent need to pass through locks. A flight of twenty or thirty locks in a stretch of a few kilometres

Iron wheels on iron tracks, first used at Coalbrookdale to transport goods down to the banks of the Severn for loading into barges, were inevitably linked with the steam engine, to create the greatest popular marvel of the Industrial Revolution – the railways. These early coal wagons are at the Beamish Open Air Museum in Durham.

was not uncommon. And then there was the question of the seasons and the state of the rivers, for without sufficient water the shallow canals were unnavigable.

By the beginning of the nineteenth century, therefore, the canal system of Britain was under great pressure to pick up speed and match the beat of the industrial engine, or fall to one side. Ironically, the efforts of the canal managers to speed up the movement of traffic spawned an innovation which in the end brought about the eclipse of the waterway system.

During the eighteenth century, as coal production increased, the mine managers had started to provide wooden tracks for their wagons, particularly from the mines down to the canals where the barges were loaded. The heavily-laden wagons were still pulled by horses, but at least they did not sink in the mud. When iron became cheaper and available in quantity, after Abraham Darby's discovery, it was natural to line the wooden tracks with iron plates – plateways, as they were known. The next step was to replace the iron-lined wooden tracks with lengths of iron rails. The final modification was to fit wagons with iron wheels to withstand the heavy wear.

Almost inevitably, this was first done at Coalbrookdale. In 1767 the first wagons in the world with cast-iron wheels ran on cast-iron rails from the Coalbrookdale Company's warehouse to the barges along the banks of the Severn. The tracks where the rails were laid have been preserved on the bank, in front of the ornate neo-Gothic warehouse, which today forms part of the Coalbrookdale 'living museum'.

With hindsight, the move from iron wheels on iron rails to the most enduring popular marvel of the Industrial Revolution, the railways, may not seem very great. Yet there were, in those opening decades of the nineteenth century, some formidable barriers still to overcome. One, strangely enough, was a stubborn conservatism on the part of the great James Watt. Having invented the separate condenser, Watt had made very little change to his engine since the patent had been granted in 1769 and then extended to 1800. In particular, he had resisted all advice to use increased steam pressure to give added power to the piston. His steam pressure was not much above atmospheric pressure and he was convinced that anything higher would be dangerous.

Among those who disagreed with Watt were a number of engineers in Cornwall, where Watt's engines were widely used because of their fuel economy. The Cornishmen were trying to build an engine without the separate condenser, on which they had to pay royalties to Watt. But without the condenser they could not get enough power – unless they could somehow increase the steam pressure on the piston.

One of those engineers was Richard Trevithick. He was the son of a mine captain, and while at school had been described as a 'disobedient, slow, obstinate, spoiled boy, frequently absent and very inattentive'. But although Trevithick never received any formal engineering training, he learned a lot as he wandered around the mines. He saw Boulton and Watt engines work and felt that far more power could be obtained by using more steam pressure. In 1797, when he was twenty-six, Trevithick built a model of a high-pressure steam engine.

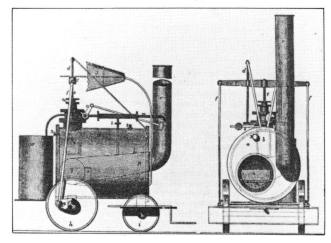

The forerunner of the railway engine was the steam road carriage built by the Cornish engineer, Richard Trevithick, and first demonstrated publicly in 1801. *Trevithick's carriage later caught fire and was destroyed and no contemporary illustration of it survives. This later drawing is in the Science Museum in London.*

The locomotive built at Coalbrookdale to Trevithick's design and driven on a colliery line in Wales in 1804 was the world's first steam locomotive. *This conjectural model is in the Science Museum in London.*

In Trevithick's engine the piston was forced down solely by the pressure of the steam fed into the cylinder. He circumvented Watt's patented condenser by exhausting the steam straight into the atmosphere from the cylinder instead of condensing it. Trevithick also got rid of the overhead beam, and connected the piston directly to a large rotating fly-wheel by a crank. All these advances produced a steam engine that for the first time was light enough and compact enough to mount on wheels – which is what Trevithick proceeded to do.

Although Trevithick almost certainly did not know it, a French inventor, Cugnot, had already built the world's first self-propelled steam vehicle in 1769. It apparently did not perform well and nothing more was ever heard of it. Trevithick went on to build and demonstrate a full-scale steam carriage of his own design. It is believed to have had a wooden body like a cart, a tall chimney above the boiler, and a single-cylinder engine driving one pair of wheels through a crank.

Trevithick was ready to prove his belief – held in considerable doubt at the time – that smooth wheels turned by an engine would have sufficient road adhesion to propel a vehicle. Since no one had ever seen a driven wheel before, he and a friend tried out the theory by propelling a one-horse chaise up various hills by turning the spoked wheels by hand.

The first run of Trevithick's steam carriage was made on Christmas Eve, 1801, near Camborne in Cornwall. According to accounts, it successfully puffed its way up a hill with several passengers aboard. Nothing survives of this historic vehicle, incidentally, because a few days later its steering gear broke down outside Camborne and, in the words of Trevithick's friend, Davies Gilbert (later President of the Royal Society), 'the carriage was forced under some shelter, and the Parties adjourned to the Hotel, and comforted their Hearts with a Roast Goose and proper drinks, when, forgetfull of the Engine, its Water boiled away, the Iron became red hot, and nothing that was combustible remained either of the Engine or the house'.

Trevithick was now convinced that high-pressure steam was the power source of the future, and that he could use it to put a steam locomotive on rails, and thus replace the horse-drawn tramways throughout Britain. First, however, he went to Coalbrookdale to carry out tests. The Coalbrookdale Company was famous for the quality of its iron boilers, and he asked them to make one to his specifications. In 1802 an admiring but somewhat apprehensive crowd stood well back as they watched the engine and boiler running at the then unheard-of steam pressure of one hundred and forty-five pounds per square inch.

From that moment on, Trevithick was kept busy supplying his high-pressure engines for many different purposes, but he kept his chief objective in view. In 1803, while he was in Wales supplying steam engines to the iron works and colliery at Penydaren, he persuaded the ironmaster, Samuel Homfray, to invest in his 'railway engine'. Homfray did so, and bet a rival ironmaster five hundred guineas that a steam locomotive could haul ten tonnes of iron on the tramway (usually horse-drawn) from Penydaren to Abercynon, a distance of just under sixteen kilometres. Trevithick went off to Coalbrookdale and had his engine built.

On 21 February, 1804, Trevithick handsomely won the bet for Samuel Homfray. His steam locomotive, puffing and hissing, pulling five wagons containing ten tonnes of iron and seventy people, trundled along the tramway at just above a fast walking pace. It covered the full distance in four hours and five minutes, stopping only for Trevithick and his passengers to cut down overhanging trees and move a few boulders near the track.

In this historic journey we can see the outline of the developments soon to follow, when in 1829 George Stephenson and his *Rocket* won the famous Rainhill Trial to choose the engine to operate the world's first passenger railway – the Liverpool & Manchester. That victory unleashed the 'railway mania' which swept Britain and spread with remarkable speed around the world.

The figure of Richard Trevithick has been overshadowed by Stephenson's deservedly popular image, but even in 1804 Trevithick's locomotive had the essential features of the railway engines which were to persist throughout the whole of the steam era. Its smooth, coupled wheels relied on friction on the rails to transmit the drive from the pistons. The high-pressure steam did not pass into a condenser after

doing its work, but was exhausted up the chimney to help the draught through the fire-box. In doing so it produced the characteristic puffing, that most evocative call-sign of the steam locomotive.[46]

The full story of the railways needs no telling here, except to recognise their significance as a milepost in the long, footsore march of the common man. The iron monsters put an end to the horse-drawn past and the tyranny of distance which had kept rural society fragmented. For the first time, for the mass of people, the daily circle of contact and acquaintance was spectacularly widened. The railways gave mobility to those whose horizons had traditionally extended little further than the bounds of their village. Among the few who remained unconvinced was the Duke of Wellington who, after attending the opening of the Liverpool & Manchester Railway, complained that the railways would merely 'enable the lower orders to go uselessly wandering about the country'.

Of course, the Industrial Revolution had its dark side, and its images of city squalor and industrial servitude are haunting, but overall there was a general and lasting improvement in living standards. In nineteenth century Britain, per capita disposable income quadrupled. A sixteen-fold increase in available goods and services transformed the very nature of poverty. The railways provided a means of quickly and

In 1825 George Stephenson's engine, 'Locomotion', pulled the first train to carry passengers as well as freight, from Stockton to Darlington in Yorkshire. *This working replica operates at the Beamish Open Air Museum in Durham.*

177

cheaply distributing the manufactured goods which were beginning to transform the home life of the working population. The railways gave Britain the momentum to cross a great divide, that watershed between the medieval world, where existence depended very much upon human effort, and the world that we know, in which our way of life depends upon machines.

From the grime and soot of nineteenth century industrial England one bold figure after another stepped forward to lead the advances on different fronts. One such figure was both engineer and artist, pragmatist and visionary – Isambard Kingdom Brunel. Perhaps the greatest expression of his daring and imagination is to be found resting in the dry dock at Bristol where he launched her in 1843, and where she is now being restored to her original splendour: the SS *Great Britain*.

'In all that constitutes an engineer in the highest sense he had no contemporary and no predecessor.' *Isambard Kingdom Brunel, one of the giant figures of the Industrial Revolution National Portrait Gallery, London.*

In building *Great Britain*, Brunel was expressing all the enthusiasms and innovative technologies of his time. He took a bold stride into the unknown and the untried. His concept was a huge assembly of concepts which had yet to be proved or even tested. As it happened, they worked superbly. *Great Britain* was to be the forerunner of all the great ocean liners of the modern era.

Brunel was a versatile, all-round engineer who had already made his reputation with the first attempt to drive a tunnel under the Thames (later completed by his father, Marc Brunel), his elegant design for the Clifton suspension bridge across the Avon Gorge (completed after his death), and the broad-gauge Great Western Railway from London to Bristol (which for decades ran the world's fastest steam trains). In shipbuilding he had also won a spectacular victory in the inaugural trans-Atlantic steamship race in 1838 with his racy paddle-wheeler, SS *Great Western*, built of oak by traditional methods and displacing some sixteen hundred tonnes. She proved herself to be the fastest ship afloat and, with sixty-seven crossings to New York in eight years, became the first holder of the Blue Riband of the Atlantic.

Encouraged by this success, Brunel threw the shipwrights' textbooks overboard for his second ship for the Great Western Steamship Company, *Great Britain*, which was laid down in 1839. Brunel had planned to build his new ship of oak in the usual way, only larger – around two thousand tonnes displacement. But when he saw a small coastal paddle steamer named *Rainbow*, which plied between Bristol and Antwerp, he

SS 'Great Britain', the first screw-driven iron-hulled ocean-going ship, set the pattern for all the great liners that were to follow her. *In favourable conditions she used her sails to assist the engines.*

changed his mind. *Rainbow's* hull was made of iron plates, and Brunel's imagination was immediately fired. With the strength and rigidity of iron he believed he could build an ocean liner such as no one had ever seen before. The company accepted his plan, and the iron hull plates and ribs were ordered from Coalbrookdale.

Brunel now increased the size of *Great Britain* dramatically – to more than three thousand tonnes. Measuring one hundred metres overall, she was twice as big as any existing ship. To make her rivetted hull rigid, Brunel introduced a new principle of shipbuilding – watertight bulkheads. The construction of her huge engines ran into a problem when it was found that no existing tilt-hammer was big enough to forge the massive iron shaft which would drive the two great paddle-wheels. In desperation, Brunel's engineer wrote to James Nasmyth, a partner in a well-known firm of machine-tool makers.

'In little more than half an hour after receiving Mr Humphrey's letter narrating his unlooked-for difficulty,' wrote Nasmyth in his autobiography, 'I had the whole contrivance, in all its executant details, before me in a page of my Scheme Book.' What Nasmyth designed that morning was the mighty steam-hammer which was to become world famous as the engineer's most formidable tool for forging large metal components.

As it happened, the original engines were scrapped even before they were completed. This abrupt decision by Brunel was caused by the appearance in Bristol in 1840 of a three-masted topsail schooner with auxiliary steam called *Archimedes*. Neither Brunel nor any of his team had ever seen anything quite like *Archimedes* – and the name is the clue. She was driven not by paddle-wheels but by a revolutionary new form of propulsion, the screw propeller.

Brunel persuaded his company to charter *Archimedes* for six months for performance tests. He then grasped the potential of this new principle fearlessly, and abandoned paddle-wheels for *Great Britain*. Instead, she was to be driven by a six-bladed screw nearly six metres across. New engines were built, with four mighty cylinders more than two metres in diameter. They were designed to turn the main shaft at a stately eighteen revolutions per minute.

SS *Great Britain* was launched by the Prince Consort in 1843 amid scenes of great excitement – gun-firing, martial music and bell-ringing all over Bristol. The port was crowded with small pleasure craft as her huge black-painted hull was eased out of the dock. She was fitted out in the Thames and again drew huge crowds. London society, led by Queen Victoria and her Consort, inspected her spacious saloons and sixty-four staterooms, all richly carpeted and panelled. Nothing like *Great Britain* had been seen before, and when she made her maiden run to New York in 1845 she began the era of the ocean liner.

The speed and seaworthiness of the world's first screw-driven ocean-going passenger ship were convincingly demonstrated on her first four trans-Atlantic crossings. On the fifth voyage, carrying one hundred and eighty passengers, the largest number ever to sail on one ship, *Great Britain* struck disaster. She ran aground on a rocky sandbank off Ireland (apparently through the use of new and inaccurate charts), and the passengers and crew had to be taken off by carts across the sand at low tide.

Brunel went over to inspect his stranded masterpiece, and after protracted efforts she was floated off and towed to Liverpool, where her hull was repaired without difficulty. Although Brunel's faith in iron ships had been vindicated, the costs ruined the Great Western Steamship Company, and both *Great Britain* and *Great Western* had to be sold. *Great Britain* was put on the Australia run by her new owners and made thirty-two voyages to Melbourne in twenty-three years. With her six masts set with sails to assist the engines she was a fine sight, running before the trade winds, and became one of the most popular ships on the Australian route. On one voyage, in 1861, she carried the first All-England cricket team to visit Australia.

During this period *Great Britain* was diverted twice, to serve as a troopship during the Crimean War and the Indian Mutiny. In 1876 she was laid up and advertised for sale. The end seemed near. But in 1882 she reappeared in the Mersey with three masts rigged for sail, her engines removed, and her iron hull sheathed in wood. In this form she made two voyages round Cape Horn to San Francisco as a freighter, carrying coal on the outward voyage and wheat on the return.

It was on the third such voyage, in 1886, that *Great Britain* ran out of luck. She was battling her way round Cape Horn with a load of coal when she ran into extremely heavy weather. Her cargo shifted and she lost her upper masts and rigging, and had to run before the gale to the Falkland Islands. Put up for sale because of the cost of repairs, she was bought by the Falkland Islands Company and used for storage until 1937, when she was finally abandoned. Her hull was beached in Sparrow Cove, and became the haunt of sea birds. SS *Great Britain*, one of the most stirring symbols of Britain's leadership of the Industrial Revolution, seemed headed for oblivion.

SS 'Great Britain' was launched in 1843, and her varied career appeared to have come to an end when she was abandoned in the Falkland Islands after a fire in 1937. In 1967, however, the old ship was salvaged and towed back to Bristol. *She is now being restored in the original dock in which she was constructed.*

In the 1950s the director of the San Francisco Maritime Museum began trying to arouse interest in a salvage operation, but nothing happened for another decade. Action finally came in a typically British fashion. A letter to *The Times* in 1967 set off a campaign to save *Great Britain*. Funds were raised and expert help recruited, and in 1970 the old ship was rescued and towed home to Bristol on a submersible pontoon. When she was floated off the pontoon and towed up the Avon on her own bottom a hundred thousand people lined the banks to greet her. She was eased into the very dock where she had been launched, and the long process of restoration began.

Today SS *Great Britain*, repaired, repainted and re-masted, is a national monument – and not only to Isambard Kingdom Brunel, of whom it was said after his death that 'in all that constitutes an engineer in the highest sense he had no contemporary and no predecessor'. This great ship also projects the vaulting imagination and spirit of confidence which was in tune with her builder and his times. As the Iron Bridge would become the bridges which spanned the world, so would the iron ship become the liners which reduced the oceans and populated new lands. They were connecting links which cut the world down to a new size, and in the process liberated millions from an ancient parochial bondage to their surroundings. These two concepts, executed on a masterful scale, shone with first magnitude in the fiery sky over Britain, lit by the furnaces of the Industrial Revolution.

Brunel's innovative use of the essential fabric of that revolution – iron – extended beyond railways and ships and bridges into other and more venturesome fields. In 1851 he designed and built one of the great new cathedrals of the railway age – Paddington Station, the London terminus of his Great Western Railway. In this concept he was

Cast iron gave architects the first new building material in thousands of years, and they used it exuberantly. *The Palm House at Kew Gardens.*

greatly influenced by Joseph Paxton's fairy-tale edifice of iron and glass for the Great Exhibition of 1851, the Crystal Palace.

'I am going to design . . . a station after my own fancy,' Brunel wrote to a friend, 'that is, with engineering roofs, etc. etc. Now, such a thing will be entirely *metal* as to all the general forms, arrangements, and design; it almost of necessity becomes an Engineering work.' In the result, Brunel's rigorous principles are concealed by artistry. Broad, graceful arches span the tracks and platforms, intersecting over the aisles and naves with geometric logic to form a vast, floating canopy of glass and cast iron. Brunel's bravura use of iron projected a growing confidence in this structural material – the first significant innovation in building materials in thousands of years.[47]

But architects and engineers were also beginning to push against the limitations of iron. Cast iron was cheap to make in blast furnaces and was useful in many structural forms, but was brittle because of its high carbon content. Wrought iron was stronger, and could be made by melting cast iron and stirring or puddling it to remove as much carbon as possible. But the process added to the cost, and the results were not always uniform because of variable carbon content. What everyone wanted was cheap iron, in quantity, with just a little carbon – between 0.1 and 0.3 per cent, evenly distributed. What, in fact, is now called mild steel. But in the middle of the nineteenth century such a steel was not available.

As already described, it was known from about 1200 BC onwards that iron could be 'steeled' by laborious work at the forge, when traces of carbon from the fire entered the surface layer of iron. By the Middle Ages there was also a method of steeling bars of iron by heating or 'roasting' them with charcoal in a closed furnace or oven. What had remained unknown in Europe until the eighteenth century, however, was the secret of making liquid steel, which had been discovered in India as early as 500 BC. This technology was re-discovered in 1740 by Benjamin Huntsman, a Yorkshire clock maker, who was dissatisfied with the steel he had to use in his clock springs. Huntsman's crucible process, which came to make Sheffield famous, was no different in principle from the making of Indian 'wootz' steel, discovered two thousand years earlier.

Huntsman's simple yet reliable method of making high quality steel can still be seen at the Abbeydale Industrial Hamlet, in the suburbs of Sheffield. This iron works started making agricultural edge tools in 1714, and in 1833 began producing them from crucible steel. Its chief products were closely related to the needs of the growing population and the demands for more food – scythes, sickles and reaping hooks. The last scythes were made in 1933, but the complex of stone and brick buildings and waterwheels has been preserved with all its technology, including furnaces, forges and water-driven trip hammers and grinding stones. While a handful of old Sheffield steel workers are still alive, the Abbeydale Hamlet still carries out the occasional 'pour' of steel.

Huntsman's process begins with pieces of high-quality wrought iron, which are mixed with charcoal in clay crucibles or pots for melting. The pots, which are like tall vases, are made at the Hamlet from clay and then baked to a cherry-red temperature to make them strong. The pots are lowered through round melting holes in the floor of the furnace room into the furnace and charged with the mixture of iron and charcoal. A lid is placed on the pots, the melting holes closed, and the furnace is fired for three or four hours, with natural draught raising the temperature to about 1550°C.

When the iron melts and the charcoal burns, carbon is evenly distributed throughout the molten metal – something which was never possible by steeling iron bars at the forge or roasting them in ovens. The crucible is then lifted out of the furnace, and the molten metal poured into a narrow mould to form a bar of steel of superlative hardness and consistency.

At Abbeydale, the bars were re-heated in the forge and fabricated into blades for farm implements. For most cutting tools, a thin strip of steel was welded between two pieces of wrought iron and the sandwich hammered into its final shape under the massive water-powered trip hammers. The final sharpening was done on water-driven grinding wheels.

Sheffield products were unmatched for their sharpness and durability, but there were severe constraints on the production of this steel. The cast bars, weighing perhaps twenty-five kilograms, were close to the limit in size that could be produced from a crucible. As an industrial material, crucible steel was far too valuable to meet the growing appetites of the Industrial Revolution. The problem of producing a cheap improvement on cast iron required a solution on a quite different scale. That solution was found in 1856 by an English engineer named Henry Bessemer.

Bessemer was seeking a method of making wrought iron from cast iron on a large and economic scale by improving on the puddling process. It was a highly desirable objective because wrought iron had resisted all attempts to mechanise its production, which was labour and fuel intensive. There were more than three thousand puddling furnaces at work in Britain but their average output was less than two tonnes a day, and all in small batches. So instead of having the melted cast iron stirred by human labour to remove the carbon, as in the puddling process, Bessemer tried blowing air into it through pipes in the bottom of a large barrel-shaped converter, open at the top.

The result was unexpectedly violent. The oxygen in the air not only removed the carbon but, by an exothermic reaction, generated heat in the molten iron instead of cooling it, as Bessemer had expected. In a matter of seconds a fountain of blazing gases, smoke and sparks roared from the mouth of the converter in a spectacular eruption.

After about ten minutes the flames died down somewhat, and Bessemer was able to get near enough to his converter to turn off the air. He tapped the iron into a mould and tested it. What he had produced was malleable or workable iron with a low carbon content – virtually wrought iron. Bessemer announced his discovery on 11 August, 1856 at a meeting of the British Association in Cheltenham. It was one of the most momentous announcements ever made before that body.

The product of Bessemer's process was not exactly the same as wrought iron. Wrought iron contained impurities in the form of threads of slag, but this helped it to resist rusting. Bessemer's 'mild steel', as it came to be called, was purer and rusted more easily. But it could be forged and rolled just like wrought iron, and its mechanical properties were better.

In addition, Bessemer's converter had two outstanding advantages – it used no fuel, only air, and was therefore cheap to operate, and it was extremely fast. A puddling furnace could make two hundred and fifty kilograms of wrought iron in about two hours. Bessemer could produce the same quantity in twenty-five minutes. More importantly, the size of the converter could be scaled up without increasing the time factor, so that several tonnes of iron could be treated in less than half an hour. Henry Bessemer had found the way to produce a metal that was just as good or better than wrought iron for almost every purpose – and it was cheap.

The advances in steel making triggered by Bessemer's initial discovery (which was matched simultaneously but independently by William Kelly in the United States) unleashed the full energies of the Industrial Revolution throughout the world. Steel became what it has remained ever since – the most important and widely-used of all metals.

Water power persisted well into the Industrial Revolution, and even beyond it. This trip-hammer, driven by a water-wheel, was used for shaping scythe blades until 1933. *Abbeydale Industrial Hamlet, Sheffield.*

185

By the end of the nineteenth century, steel had girdled the world with bridges and railways, galvanised it with machines, and changed the profile of cities. Remarkably enough, the most striking image of this transformation is to be found not in Britain, where it all began, but across the Channel in France.

When Gustave Eiffel completed his great tower in 1889 to mark the anniversary of the French Revolution, it was the focal point of the Paris World Fair. It contained more than ten thousand tonnes of iron and steel. But it was far more than the tallest structure on the face of the Earth, the most daring use of iron ever attempted. It was a gesture of continued buoyancy and confidence in the onward march of material progress – a promise, it seemed to many, of what technology could offer mankind, poised to enter the next century.

It was appropriate that Eiffel should also have constructed the steel core for that other symbol of optimism, the Statue of Liberty, which was made in sections from copper in Paris before being shipped to New York, to mark the new frontier of a new age. For it was across the Atlantic, in Thomas Jefferson's great republic, that the mood and the technology would be caught, to run away with the world.

One of the most striking images of the Industrial Revolution is to be found not in Britain, but across the Channel. *The Eiffel Tower, completed in 1889, contains 10 000 tonnes of steel and cast iron.*

CHAPTER 9
Into the machine age

On the windswept, misty clifftops of Cornwall, ghostly monuments to a once great industry loom like medieval castles. They are the ruins of the stone engine houses which stand sentinel over the abandoned tin and copper mines. With their tall chimneys and massive granite walls, built to take the strains of the beam pumps that laboured to drain the shafts, they are enduring testaments to the time when Cornwall was the mining centre of the world, and Britain was the leader of the Industrial Revolution.

But by the middle of the nineteenth century the mines of Cornwall had gone just about as far as they could go. The working faces were so deep and distant from the surface that the cost of mining was becoming prohibitive. This handicap was to prove fatal when great new copper and tin provinces were opened up – largely by Cornish miners – in distant corners of the world. In much the same way, the skills and innovations of the British, who had led the Industrial Revolution, eventually brought into existence rivals to Britain's ascendancy – rivals which in time began to dwarf her.

The 'Great Exhibition of the Works of Industry of all Nations', which opened in London in May, 1851, was designed as a grand showcase of British achievement. It was a bold statement of confidence and pride in one hundred and fifty years of technological progress. The building in Hyde Park which housed it, the Crystal Palace ('a rare pavilion, such as man saw never since mankind began'), could not even have been erected fifty years earlier, since it was made entirely of glass and pre-fabricated interchangeable iron parts – products of the new machine-tool industry. And although more than one hundred thousand exhibits from all parts of the world were on display in four 'Grand Departments' – raw materials, machinery and mechanical inventions, manufactures, and sculpture and plastic art – more than half the total space was taken up by Britain and her colonies.

In his 'Exhibition Ode', published in *The Times* on opening day, William Thackeray directed properly admiring attention to the British products:

> Look yonder where the engines toil:
> The trophies of her bloodless war:
> Brave weapons these.
> Victorious over wave and soil,
> With these she sails, she weaves, she toils,
> Pierces the everlasting hills
> and spans the seas.

There are many historians who now see this display as marking not only the climax but also the zenith of Britain's pre-eminence as the world's first industrial power. 'Whatever the degree of Britain's industrial supremacy at the time of the Great Exhibition, doubts were already being expressed as to whether it could be maintained,' observes Asa Briggs.[48]

The reasons lay deep in the structure of British society. Some were to do with a seemingly ineradicable resistance to change. James Nasmyth, the engineer whose inventions included the steam hammer, wrote that he observed 'a certain degree of timidity' among British mechanics, resulting from 'traditional notions and attachments to old systems'. An American commentator noted in 1851 how 'English overseers are trained too much to one thing or machine and do not adapt themselves readily to circumstances, finding everything wrong which they have not been accustomed to'. Such an inclination was not restricted to overseers only – electrical machines had been left out of the Great Exhibition on Isambard Kingdom Brunel's advice that they were 'mere toys'.

There was another and less obvious resistance, the insidious effects of which can be traced even into Britain's contemporary industrial decline. That was the way the governing class in Britain turned its face away from the realities and demands of a manufacturing economy and denied the necessities to support it. Self-made industrialists did not want their children to go back into the works but to the emerging Victorian public schools, alongside the children of the pre-industrial elite, the aristocracy and the landed gentry.

C. P. Snow put it bluntly in his essay 'The Two Cultures': 'Almost none of the talent of the elite, almost none of the imaginative energy, went back into the revolution which was producing the wealth. The traditional culture became more abstracted from it as it became more wealthy, trained its young men for administration, for the Indian Empire, for the purpose of perpetuating the culture itself, but never in any circumstance to equip them to understand the revolution or take part in it. Farsighted men were beginning to see, before the beginning of the nineteenth century, that in order to go on producing wealth the nation needed to train some of its bright minds in science, particularly in applied science. No one listened.'[49]

In 1851 the cloud on the horizon was still small enough to ignore.

After all, Britain was still ahead of all other nations in the transformation from a rural to a manufacturing economy. The proportion of her total working population in agriculture was down to a quarter. Belgium, the next most intensively developed country, still had half her workers on the land. Britain's chief export earners, cotton goods, were equal in value to those of all of Europe combined. Yet the warning signs were there. They were coming not from Britain's traditional and historic rivals on the Continent, but from a new quarter – from across the Atlantic.

The Americans had sent exhibits to the Great Exhibition, and although they were vastly outnumbered by those from Britain and the Empire they made quite an impression in the section devoted to agricultural and horticultural machines and implements. One American observer even wrote that 'to Americanisc British land and sea machinery would half revolutionise Britain'.

By thc time the Great Exhibition closed in October 1851, more than six million people had passed through the Crystal Palace (which was later moved to South London, where it survived until destroyed by fire in 1936). Few of those proud citizens knew of the wave of energy and enterprise which was beginning to gather on the other side of the Atlantic, and which within two decades would take Britain's former colonies surging past the mother country on their way to unimagined levels of productivity and material wealth.

The Statue of Liberty, the symbol of the New World's welcome to the 'tired and poor' of the Old, was a gift from France to the new Republic, where the Industrial Revolution would be taken up and transformed. *The statue was first assembled in Paris from sheets of copper bolted to an iron core, designed by Gustav Eiffel, and then dismantled and shipped to the United States in 1884 for erection on its stone plinth at the entrance to New York harbour. Its correct name is 'Liberty Enlightening the World'.*

Already, by 1851, America had passed through its early phase of industrialisation, based on water power. It now had huge steam-driven textile mills in New England, rivalling Manchester's in productivity and technology. Pittsburgh's output of iron goods was doubling every few years, and by 1870 it would move almost wholly into steel production, whereas barely ten per cent of Britain's iron output would be in the form of steel. Steamboats were plying busily up and down the Mississippi with goods and passengers. And there was already twice the length of railway tracks in the United States as in the whole of Europe, including Britain.

By the time the Statue of Liberty was shipped from France and erected at the entrance to New York Harbour in 1866, to give her welcome to 'the tired and the poor' from the Old World, the Americans had already surpassed the British and the Europeans in the application of science and technology. Henceforth the old nations of Europe would no longer show the way, but rather accompany the United States into the Age of Machines. The New World was about to re-define the life of the common man.

The first ironworks on the American continent, at Saugus in Massachusetts, raised a question which has fascinated and challenged the rest of the world for the past hundred and fifty years: how were the Americans able to take the technology and the methods of the Old World and transform them in such an extraordinary way that they could shape an entirely new way of life, first for themselves, and then for the other developed nations of the world?

The Saugus ironworks is on the northern outskirts of Boston. In 1646, when the furnace and its ancillary works were built on the banks of the Saugus River, it was close to the Massachusetts Bay colony, one of the major settlements of the original Puritan migrations from England. The ironworks was established to meet a problem which had begun to impose severe constraints on the colonists after the first large immigrations ended in the 1630s – a critical shortage of iron.

At that time the Massachusetts Bay colony numbered about twenty-five thousand people, ninety-five per cent of whom were small farmers or tradesmen. There was an expanding demand for tools and agricultural implements, for nails and wire and bolts, for pots and household utensils. But as the number of ships arriving from Britain became fewer, iron products became scarcer and more expensive. The colonial government in 1641 enacted a law encouraging 'the discovery of mines', but the thick forests along the eastern seaboard of North America initially hampered exploration. No less well-informed an authority than Benjamin Franklin was to conclude that North America possessed 'no useful ores' of desirable metals.

Even when deposits of 'bog iron' were found in swamps, there was no capital and no skilled workforce available to exploit them. In 1641 John Winthrop Jnr, the son of the colony's governor, sailed for England to obtain investors and workers. He found a receptive industry, frustrated by the shortage of charcoal for smelting in Britain, and looking for new locations for iron making. The English ironmasters had just had

The first iron works on the American continent was built at Saugus in Massachusetts in 1646. It was a transplant from England, and did not survive very long, but it trained the first generation of American ironmasters. *The Saugus iron works on the outskirts of Boston has been fully restored, with all its water-wheels in working order.*

most of their works in Ireland burned down by the Irish, who complained of the exploitation of their forests.

Winthrop returned to Massachusetts with investment funds and skilled workers and set up an iron-making company on the Saugus River, where there was adequate water for power, unlimited hardwood forests for charcoal and deposits of bog iron close by. There was a blast furnace to make cast iron, blown by large leather bellows driven by waterwheels. Some of the iron was cast directly into utensils and pots, but most of it was taken to the forge to be converted into wrought iron. It was melted in the 'finery' and then beaten under the heavy water-powered trip hammer. The bars were finally reduced to usable iron strips and rods in the slitting mill.

As it turned out, the Saugus ironworks did not operate for many years. The difficulties of a new enterprise in a remote colony and the comparatively high operating costs made its prices less competitive, once shipments from Europe began to flow in again. But the men who had worked there went out and helped to set up furnaces and forges in the other Atlantic colonies, and their sons became the first generation of American ironmasters.

The Saugus ironworks complex has been completely reconstructed, in working order, on its original site on the Saugus River. Like many of the names along the river – Hammersmith and Bristol and Braintree – it was a transplant from England. In that sense, it marked the first introduction of the Industrial Revolution to America. But in the style of its log-wall construction, and in the particular things that if made, it was evolving a style which was already identifiably American. For the Americans would proceed to do with the Industrial Revolution over the next two hundred years essentially what they did with all the citizens of all the nations who flocked to the New World as immigrants – they naturalised it, and they made of it something specifically American. Saugus was thus the working prototype, the forerunner, of that immense, phenomenal growth of productive industry which is still unmatched in history, and which provided the framework for the new American civilisation, and was the source of its dynamic energies.

There was another factor, which was to be all-pervasive in the evolution of that unique style of existence, the 'American way of life'. What set the Americans apart was their devotion to the ideal of work and individual achievement. In the words of one historian, 'they all devoted themselves to it, all of the time'.

It was not just the opportunities offered by the luxury of space, of room to move and, later, by the treasure-house of resources. It stemmed in some essential manner from their background of social, religious and economic constraint in the Old World, and the promise of their arrival in the New. The settlement of North America was a fresh start, as well as a continuity. The new arrivals were competent and full of expectation. They set out on a revolutionary quest for the good things in life, which they were prepared to achieve by work rather than by privilege or paternalism. Where there was no inherited wealth and few material possessions, everyone worked. Work was not considered to be degrading or humiliating. To work for profit was an admirable aim in life, and all the prejudices were for it and not against it. And what developed from this – the American way of doing things – has largely determined the way we live today.

In the late 1700s all this was still to be realised, or even conceived. In the wake of the crippling War of Independence, which ended in 1781, the new nation felt the constraints of resource shortages even more keenly. One of the most critical shortages was manpower. There were just not enough skilled men and women to meet the demand. And so the cry went up for machines, and the Americans began that remarkable stampede to invest in machinery which was to change the history of the world.

Eli Whitney developed the cotton 'gin' (short for engine), which separated seeds from cotton fifty times faster than hand sorting. Cyrus McCormick ended thousands of years of reaping by scythe and hook when he invented the horse-drawn mechanical harvester. Robert Fulton's steamboats conquered the Mississippi. The cotton mills and factories were filling up with standardised machines, with interchangeable parts and a deliberately high rate of obsolescence. Most significantly, the workers did not fight such introductions. Change was the

American way of doing things. An English Parliamentary report observed with incredulity that 'American workmen hail with satisfaction all mechanical improvements, the importance and value of which, as releasing them from the drudgery of unskilled labour, they are enabled by education to understand and appreciate'.

Such industrial growth placed increasingly heavy demands upon the mineral resources of the Atlantic seaboard states. There were scores of coal, iron and copper mines in operation in Pennsylvania, New York, New Jersey and Virginia, but the deposits were comparatively small, and their output was inadequate for either needs or ambitions in the new nation. The hunger for resources turned eyes to the vast, untapped expanse that lay to the west.

In the seventeenth and eighteenth centuries French fur traders and Jesuit missionaries, pushing inland into the northern wilderness, had brought back tales of extraordinary riches along the shores of the Great Lakes, in what is now Michigan. They spoke of large lumps of copper lying on the ground, or exposed in shallow holes. The most talked-about find was a copper boulder of almost pure metal weighing one and a half tonnes, which lay half exposed near the Ontonagon River. (The 'Ontonagon Boulder' was eventually bought from the Chippewa Indians for $150, and after passing through the hands of various prospectors and companies was confiscated by the United States government and presented to the Smithsonian Institution in Washington, where it is still on display.)

The French and later the British governments had encouraged companies to prospect around the Great Lakes but nothing came of those early sorties. It was not until 1840, in the new mood of adventure and mineral hunger in the United States, that rumours became reality. A geologist, Douglass Houghton, went ashore by canoe on the Keweenaw Peninsula in Michigan and found copper, as predicted, literally lying everywhere.

Houghton's report, published in 1841, encouraged a few adventurous prospectors to seek out the area, but it was not until the US government had signed a treaty with the Chippewa Indians – this was the country of Hiawatha – that exploration began in earnest. One of the first companies to take samples from the Keweenaw Peninsula published an assay figure showing copper and silver averaging $3000 to the tonne of ore.

There was a stampede of fortune-seekers from the east to the wild, overgrown and totally unexplored peninsula on Lake Superior. To their astonishment, they found that they were not the first to seek the metal that gleamed from beneath the roots of trees and from almost every boulder they turned over. Hundreds of shallow trench mines were discovered beneath drifts of fallen leaves, some still containing stone hammers and lumps of copper. Others had obviously mined here extensively before the new arrivals – not in the time of the French or English explorations, but thousands of years earlier. What the prospectors had stumbled across was the Old Copper Culture of the Great Lakes, a truly remarkable episode in the history of metallurgy.

The coming of man to the American continent, once thought to have been fairly firmly dated to the end of the last Ice Age, has in recent decades become the subject of much debate among anthropologists. According to the long-held view, Siberian hunters moved across the Bering Strait from Asia into Alaska at the height of the Ice Age, some thirty thousand years ago, when sea levels were at their lowest and the Bering Strait was dry land. For a long time, it was thought, they were penned into the glacier-free Alaskan region by the vast ice sheet which covered Canada. Then the climate began to warm up. The ice retreated far enough to open up a clear corridor to the south, east of the Rockies, and about twelve thousand years ago the Siberian hunters poured through on to the open game lands of the American plains, teeming with wildlife. Within about fifteen hundred years man had penetrated to the very tip of South America, and populated the entire continent.

Recent archaeological finds in various parts of the United States and Mexico have challenged this view, by identifying occupation sites which are claimed to be datable to at least thirty thousand years ago. The implication that man somehow got into the North American heartland before the opening up of the Canadian ice corridor is the subject of increasing debate.

What is beyond argument, however, is that the Americas were settled long before metals came into use anywhere in the world. The likelihood of such technology reaching North America after the rise in sea level between nine and six thousand years ago is extremely small. The conclusion must be that the inhabitants of North America made their own independent discovery of metals, and that one of the earliest places where this occurred was around the Great Lakes. Here again there is dispute over dates, but they range from 1000 BC to as early as 5000 BC.

Those first metal workers were assisted by an unusual combination of geological and chemical conditions, which had caused copper to be concentrated in the rocks of the Keweenaw Peninsula and nearby islands in virtually pure metallic form. The copper had originally been deposited from solution in holes and pockets in the porous rock, created by steam and gases during the outpouring of lava which formed the landscape. Unlike the normal sequence of chemical transformation which takes place in mineral deposits, the copper here did not oxidise or join with other elements such as sulphur, but remained in its pure metallic state, contaminated only by traces of silver. The metal lay in veins between layers of rock, and in places had formed massive nuggets, some weighing hundreds of kilograms.

The ancient miners recovered the copper either by pounding the veins with stone hammers until pieces of metal broke off, or by lighting fires against the rock and dashing it with cold water from the lake to crack it. The copper had been hammered into all kinds of tools and implements, including knives, arrow heads, axes, chisels and hooks. There are charcoal remains and evidence of some heating of the metal to make it easier to work, but no suggestion that temperatures sufficient to melt copper were ever reached.

What is astonishing is the scale of these ancient mining operations. More than ten thousand separate mines have been identified in the copper country around Lake Superior. Some of the trenches in hard rock are twenty metres long and five to ten metres deep. Engineers at the Michigan Technological University at Houghton, near Keweenaw, estimate that it would have taken a thousand miners more than a thousand years to have carried out the work already discovered, and that as much as two thousand tonnes of copper may have been recovered before the arrival of the white prospectors.

Despite the removal and dispersal of untold quantities of stone tools by later mining operations, enough items remained to make an overwhelming impression on the archaeologists who began to study the area towards the end of the nineteenth century. From one set of ancient workings, at McCargoe Cove, they collected more than one thousand tonnes of stone hammers. These mining tools weighed from two to five kilograms and consisted of smooth, wave-washed stones from the lake shore. Some were grooved to take a handle, while others were simply hand-held.

The questions which arise are intriguing. Who were the people responsible for this immense effort? What did they do with all that copper? And why did they apparently abandon the mines long before the present tribes arrived on the scene? The Chippewa Indians, when first encountered by the French, said that they knew nothing of the copper mines.

The identity of the early copper miners is a complete mystery because no burials, habitations, identifiable pottery or other cultural clues have ever been found in the area – nothing but mining tools and worked copper articles. The inference is that they came from somewhere else to mine the copper, perhaps during the summer, because in winter the land is deep in snow. There is a possibility that they were connected with the 'mound builders' – the people who created large earth burial mounds all across North America in the first millennium BC. Many copper articles have been found among artefacts in these mounds, but they might well have been obtained by trade.

The apparently abrupt cessation of mining on the Keweenaw Peninsula is equally puzzling. The manner in which tools, baskets of copper and half-finished artefacts were left in and near the mines suggests that those responsible intended to return. But they never did. This abandonment appears to have taken place at least five hundred years before the coming of the first Europeans, a theory which is based on the age of trees found growing in mine trenches in the 1840s.[50]

At first, the American prospectors followed the lead of the ancient miners, and extended or deepened their pits. They found copper, but it soon became obvious that the surface outcrops had been well worked over. The first significant new discovery was made on the hunch of an amateur prospector, a pharmacist from Pittsburgh named John Hays, who had cashed in his savings and joined the rush.

Hays reached the 'instant township' of Copper Harbour on the Keweenaw Peninsula in 1843. Since he knew nothing of mining, he recruited a small gang of men to work for him. He and his men tried several ancient mining areas near the township without luck. He then decided to try somewhere that had not been worked, and turned his attention to the face of a high lava cliff, where a small outcrop of copper was exposed. At the base of the cliff Hays directed his men to drive a tunnel to follow the thin vein of copper back into the hillside. They had dug in no more than about twenty metres when they encountered a solid mass of copper so big that they could not get it out through their tunnel. They had to cut it up with hammers and chisels.

The Cliff Mine, as it became known, marked the beginning of the first great mining boom in America's history. Even bigger and more successful mines soon followed, as shafts and tunnels were driven into the copper-rich spine of the Keweenaw Peninsula. And with the need to go deeper, Cornish miners arrived on the scene, bringing their good humour, their songs and customs and their unmatched skills in hard-rock mining. Leaving their own worked-out mines behind, they came here to the new abundance of copper. Around Lake Superior they dug out the foundations of the American mining industry that was soon to explode across the continent and unearth the greatest treasure trove of minerals ever known.

The great, now derelict mines of the Lake copper period – the Cliff, the Central, the Quincy, the Minesota, the Calumet and Hecla – produced between them millions of tonnes of copper and paid billions of dollars in dividends. The Michigan copper boom fed more wealth into the American economy than the California goldrush. Mining technology was pushed to new dimensions. The No. 2 shaft of the Quincy mine plunged more than three thousand metres on a steep incline, and hauling the ore from its depths required the largest steam hoist ever built, weighing eight hundred tonnes.

Some of the incidental finds were incredible, by ordinary mining standards. In 1856 a tunnel in the Minesota (the name was misspelled in the original application) struck a mass of solid copper weighing more than five hundred tonnes. It was sixteen metres long and averaged nearly four metres thick, and it took the miners more than a year to cut the pure, malleable metal into pieces small enough to be taken to the surface.

The Lake copper boom came at precisely the right time for the nation. In 1844 Samuel Morse had demonstrated his first telegraph line, between the Capitol in Washington and Baltimore (opening it with the suitably portentous message: 'What God hath wrought'). Now there was an ever-expanding web of humming copper wires spreading across the United States, the staccato, clicking messages carrying news, stock market prices and business intelligence from the booming east coast to the new frontier on the Pacific.

The success around Lake Superior did more than stimulate the economy. It reinforced a fundamental compulsion of American life – the movement to the open West. It launched the Americans on a trajectory which in little more than three hundred years was to take them from colonial status to world primacy; from their log cabins in the wilderness to walk on the Moon. In that tremendous burst of exuberance and vitality one unfailing ingredient was the 'frontier spirit' – the beckoning encouragement of that line advancing across the continent, the constantly moving frontier. The prospect of so much free space, lavishly

The first great mining boom in America took place along the shores of Lake Superior, where copper was found in huge deposits of almost pure metal. The Cornish miners who brought their skills to Michigan also brought their characteristic stone buildings and engine houses. *Abandoned copper mines on the Keweenaw Peninsula.*

impregnated with mineral treasure, provided a constant renewal of energies which might have otherwise flagged.

Unfortunately, there was an unacceptable side to this exuberant conquest of the wilderness – the destruction of Indian society and culture. Although the Americans, unlike the Spaniards in the New World, were less concerned to vanquish and 'save' nations than to master the resources of their new domain, and although they were seeking rewards under an individual democracy rather than plundering on the orders of some distant king, they were still unable to find a way of accommodating the existence and the rights of the original inhabitants of the continent. It was a failure of morality and political judgement that was to haunt nearly all European colonial expansion in the eighteenth and nineteenth centuries.

In that portrait of the early American colonial period which has been enshrined in history as well as myth, a recurring image is of a God-fearing, hard-working people planting crops and hewing wood, obeying the Scriptural injunction not to idle in the Garden of Eden but to till it and keep it. But while cultivation offered a sure return, it was slow. As the American brand of civilisation developed, so did the pace. It took too long to build a house of stone or brick, so there evolved the frame house, in which exterior and interior planks were nailed to a swiftly-erected skeleton of light timber. Laying proper ballast under railway tracks was too slow, so locomotives were made 'loose-jointed as a basket', so they could adjust to undulations in the lines.

Human energies were diverted away from agriculture and towards manufacturing. Industrial development had begun to govern both wealth and power in the world, and in America it was enshrined as the crucial objective. When this was allied with 'Yankee know-how' concerning machines and gadgets, and a workforce eager to share the rewards, it set in motion an irresistible surge of industrialisation which rapidly outpaced the Industrial Revolution in Europe.

To meet the needs of industry there had to be vast new resources of minerals – iron, copper, lead, zinc and tin to make the machines and their products; coal to fire the furnaces and steel mills and drive the endless train-loads of raw materials and goods through the arteries of the economy; and mountains of gold and silver to finance it. So a powerful incentive was added to the natural spirit of exploration along the frontier. The fortunes made in the Lake copper boom had shown the possibilities. Where would the next bonanza be?[51]

In fact it came even as the first copper miners were descending upon the Keweenaw Peninsula. Further west into the Michigan wilderness, along the remote shores of Lake Superior, a surveyor named Burt noticed erratic swings of his compass needle. He knew that iron could have such an effect, and hired a local Indian to help him look for possible outcrops of iron ore. On 23 July, 1845, Burt saw ahead of him, through the trees, what he later described as 'a mountain of solid iron ore, one hundred and fifty feet high, its crimson hue made brighter by the stark white bark of the birch trees growing on its slopes'.

It was the first find of what turned out to be one of the world's largest deposits of iron, extending around the western shores of Lake Superior into Minnesota and across into Wisconsin. The ore body lay horizontally, rarely more than about sixty metres below the surface, and was comparatively easy to uncover and dig out. In places the ore was so rich – more than sixty per cent iron – that it could be tipped straight into the blast furnaces.

This great iron province, with the incredibly rich Mesabi Range at its heart, was to become one of the most productive ever known. Output eventually reached seventy million tonnes of ore a year. The extent of the open-cut mines, which left vast shallow craters up to three kilometres across, are a measure of the growth of the American steel industry towards the end of the nineteenth century. Lake Superior iron ore made the United States a great military power.

At first, however, the development of this enormously rich natural resource was held up by a seemingly insuperable obstacle: the problem of getting the ore out. Access by land was barred by dense forests and mountain ranges. Water transport was the only possibility, but the Great Lakes were frozen over for half the year. And shipping faced an even more formidable obstacle – there was no passage out of Lake Superior to the other lakes, which had rail connections to eastern steel mills.

When the great North American ice sheet had retreated, the removal of its enormous mass was followed by a marked rise in the Earth's crust in the region now covering the north-eastern United States and eastern Canada. This had emptied a huge basin containing the Champlain Sea, and created the chain of Great Lakes with steps or sills between them, all draining into the St Lawrence river and thence into the Atlantic. In some places the difference in level between the lakes was spectacular, as at Niagara Falls. Between Lake Superior and its neighbour to the east,

Lake Huron, the fall was only about seven metres, but even this stretch of rapids was an impassable barrier to shipping.

At first, small craft loaded with iron ore were rolled on 'greaseways' around the rapids, but larger barges still had to be unloaded and reloaded. In 1851 a tramway was built to speed up the trans-shipment of the ore, but this soon became inadequate to handle the increased shipments. There was only one solution, and it was begun in 1853 and completed two years later. The Sault Saint Marie Canal outflanked the rapids which had blocked the exit from Lake Superior. Bulk carriers, which eventually grew to more than three hundred metres in length, with a capacity of forty thousand tonnes of iron ore, began to pour through the locks and across the other lakes to the railheads.

Although the Sault Saint Marie Canal, at just under five kilometres in length, ranks only twentieth among ship canals of the world, it was one of the most significant artificial waterways ever built. It uncorked Lake Superior, and opened the way for billions of tonnes of iron ore to be brought within reach of the energy stored in America's gigantic eastern coalfields. These two great resources were harnessed together to create the industrial heartland of the United States and its capital – Pittsburgh, Pennsylvania.

Where colonial America had adopted European iron-making technology virtually unchanged, the United States bent the production of steel to its own demands and its own will, and did so on an overwhelming scale. Along the banks of the three rivers that met in Pittsburgh, endless rows of blast furnaces, steel plants and rolling mills spawned beneath a canopy of smoke and steam. Iron and steel poured out in a mounting flood. Within a few decades the United States was producing more steel than Britain, Germany and the rest of Europe combined.

This is not the place to explore the many ways in which the develop-

The great Chicago fire of 1871, which cleared large areas of the city, opened the way for adventurous architects to re-build in new and enterprising ways. This was the breeding ground of the skyscraper. *National Archives, Washington.*

ing steel technology of the United States in the late nineteenth century affected the course of events in North America and elsewhere, but in one particular direction it had an unforgettable impact, and that was on the profile of the world's cities. The steel-framed tower building is the most emphatic American statement in architecture.

The skyscraper was born not, as might be thought, in New York, but in Chicago, in the aftermath of the great fire of 1871. The wide area of the city left bare by this disaster offered rare opportunities to rebuild in new and more space-effective ways. One young architect to accept the challenge was William Jenney. He boldly discarded the principle of load-bearing walls, which had always imposed limits on the heights of buildings, and erected a steel skeleton to take all the structural stresses. His ten-storey office block for the Home Insurance Company, completed in 1885, was the world's first skyscraper. It was so admired that two additional storeys were later added.

The spread of the skyscraper around the world is the most conspicuous architectural development of the twentieth century. Although these vertical ant-heaps dwarf their inhabitants and are considered soulless by many, the pressure of rising urban populations on available space has assured their proliferation. Meanwhile, the use of metal in skyscrapers has gone far beyond Jenney's original concept, and today's lofty towers are not only supported by steel but are often entirely sheathed in copper or aluminium.

In another direction, the American iron industry in the nineteenth century played a key role in the unification of the nation. The popular excitement as that possibility neared fulfilment was caught on 1 March, 1854 by the *Rock Island Advertiser*, a newspaper in a small river-boat port on the upper Mississippi, west of Chicago:

'Amid the acclamations of a multitude that no man could number, and the roar of artillery, making the very heavens tremble, punctual to the moment, the Iron Horse appeared in sight, rolling with a slow yet mighty motion to the depot. After him followed a train of six passenger cars crammed to the utmost with proud and joyful guests, with waving flags and handkerchiefs, and whose glad voices re-echoed back the roar of greeting with which they were received. Then came another locomotive and train of five passenger cars, equally crowded and decorated. This splendid pageant came to a stop in front of the depot, and the united cheers of the whole proclaimed to the world that the end was attained, and the Chicago and Rock Island Railway was opened through for travel and business.'

It was barely twenty-five years since the first locomotive had arrived in New York from Britain in 1829. It had been eagerly inspected, jacked up on blocks, wheels spinning briskly but futilely under steam, because there were no tracks for it to run on. Within a year the Americans were building their own steam locomotives, and tracks were snaking out across the land like iron tentacles.

The extension of the railroad westwards from Chicago to the Mississippi, the natural frontier to the wilderness, was the first step towards the conquest of the vastness that lay beyond. As the American author Dee Brown describes that occasion, the watchers knew that

THE CHICAGO BUILDING OF | THE HOME INSURANCE CO.

OF NEW YORK

'only the demonic power of the Iron Horse and its bands of iron track could conquer the West . . . the presence of Locomotive No. 10 on the banks of the great river that day portended more than any man or woman there could have dreamed of. Its Cyclopean eye faced westward to the undulating land that flowed into a grassed horizon and curved over the limitless expanse of the Great Plains a thousand miles to a harsh upthrust of Shining Mountains, the Rockies, and then down to the Western Sea – that goal of European wanderers and seekers across North America for three centuries, the great ocean that Ferdinand Magellan named the Pacific'.[52]

The laying of the network of iron tracks that spanned the United States in the next decade – made possible only by the stream of rails that poured from the rolling mills in Pennsylvania and Illinois – was one of the great epics of human endeavour, a lurid tale of endurance, tenacity, enterprise, greed, exploitation and opportunism without equal in modern times. Its natural climax was the 'Great Railway Race' to close the last gap between the Atlantic and the Pacific.

The world's first steel-framed office building was erected in Chicago in 1885. It was the forerunner of all the skyscrapers and towers that were to follow – even the Renaissance Centre in Detroit. *Left: Chicago Historical Society.*

By 1867 the Union Pacific track-layers, heading west, were working across the Great Plains of Nebraska towards Salt Lake City. Samuel Bowles, the much-travelled publisher of the *Springfield Republican*, recorded the name attached to those rough and temporary communities at the head of the ever-advancing line: 'These settlements were of the most perishable materials – canvas tents, plain board shanties and turf-hovels – pulled down and sent forward for a new career, or deserted as worthless, at every grand movement of the Railroad company . . . restaurant- and saloon-keepers, gamblers, desperadoes of every grade, the vilest of men and women made up this "Hell on Wheels", as it was most aptly termed.'[53]

At the same time, the rival, predominantly Chinese, Central Pacific crews were battling their way eastwards across the high Sierras from San Francisco, boring long tunnels through the solid rock while fighting blizzards and dodging avalanches. To keep the tracks open to the summit tunnels, a Scottish bridge-builder named Arthur Brown was hired to construct sixty-four kilometres of snowsheds over the line. In doing it, Brown used sixty-five million feet of timber and nine hundred tonnes of bolts and spikes. One of his biggest problems were the frequent fires in the snowsheds caused by sparks from the engines.

Desperate to reach Salt Lake City ahead of Union Pacific, and worried at the slow pace of the Sierra tunnelling, Central Pacific's manager left the tunnellers at work and took three thousand Chinese over the summit to start building the line down the eastern slope towards Nevada. In mid-winter, they dismantled three locomotives and forty cars and dragged them over the ice-clad summit on sledges and down the other side, slipping and sliding in the deep snow like Hannibal's elephants crossing the Alps. Behind them came a cavalcade of sledges and wagons, manhandled by Chinese workers through the snow, loaded with ploughs, tools, sawmills, food and enough iron rails and spikes for eighty kilometres of track.

With the spring of 1868 the race really heated up. Both Union Pacific and Central Pacific crews were laying better than two kilometres of track a day as they closed the fifteen hundred-kilometre gap. In April, the Union Pacific drove its track over the Black Hills of Wyoming at an elevation of two thousand six hundred metres. A thousand track-layers carried rails through snow behind the thousand graders levelling the rough terrain. At the time it was the highest railroad ever built. In Nevada, to push the Central Pacific line across eight hundred kilometres of barren, treeless desert, all timber and water had to be railed in from the Sierras, far behind.

As the weary track-layers came within six hundred kilometres of meeting, the grading teams far out in front found themselves grading track-bed and building trestle bridges alongside one another but in opposite directions. This wasteful and costly duplication continued for months, since each rail company was receiving government bonuses for every mile of track completed. They were finally ordered by the president-to-be, General Grant, to agree on a meeting point, and Promontory Point in Utah, north of Salt Lake City, was chosen.

In the final frenzied leg, Central Pacific's manager made a $10,000 bet

with his rival that his crew could lay ten miles (sixteen kilometres) of track in a single twelve-hour shift. The day was declared a local holiday and thousands came to watch the dawn start. The eight key men in the Central Pacific team, the rail carriers – all Irish – started at the run. As the rails were dragged by Chinese support crews from the flat-cars coming up behind, the rail carriers raced them forward into place for the gaugers, spikers and bolters. By noon they were well past half way, and had time for lunch. After twelve hours they were comfortably over the ten-mile mark, and the bet was won and distributed. On that day each rail-carrier had lifted one hundred tonnes of iron and the layers had spiked 3520 rails to 25,800 sleepers. And the Central Pacific was first into Promontory Point, a windswept valley between low hills, with the Union Pacific tracks appearing from the east a week later.

The historic link-up took place on 10 May 1869. America had hardly recovered from the Civil War, in which the railways, by speeding up the movement of the Union armies, had played an important role in the North's victory. When the speeches began, much was made of the driving of the final spike as a symbol of a union re-found, and a new point of departure. Promontory Point was a remote place and the crowd numbered only a few hundred, but the whole of America was there in spirit that day, awaiting the symbolic consolidation of the nation with iron rails.

Around noon the last rail was laid and spiked down, except for two holes. A representative of Central Pacific handed the president of Union Pacific two solid gold spikes. As they were placed in their holes, a telegraph operator at a key strung from a wire along the track tapped out a stand-by. All across the continent lines were cleared, and the

On May 10, 1869, the Great Railway Race came to a triumphant finish when the tracks of the Union Pacific from the east and the Central Pacific from the west met at Promontory Point, Utah, and two golden spikes were driven to link the nation. *The meeting point is now the Golden Spike National Park, and replicas of the two locomotives which met head to head that day now re-create the historic link-up.*

people of America waited expectantly. At 12.47 the silver sledge-hammer swung twice and the Morse message flashed across the nation: 'Done!' Amid scenes of hysteria, the lead engines of the two railroads steamed slowly up to the meeting point, head to head, and clanged triumphantly together. The transcontinental link was ready for the Iron Horses to roll, from the Atlantic to the Pacific, and before long from Canada to Mexico.

But even amid the celebrations, the reign of coal and steam, and their monarch, the railways, was already measured. Their usurper was quietly making its presence known in shady streams in far-off Pennsylvania.

The first settlers in the valleys of northern Pennsylvania had long been aware of iridescent gas bubbles rising in pools, and of a sticky black substance which sometimes collected like tar along the stream banks. It was known locally as 'rock oil', and it was used, if at all, as a kind of patent medicine. By 1850 a Pittsburgh businessman succeeded in distilling from it a fairly useful oil for lamps. It gave a better light than whale oil or candles, which was all that was available at the time.

The black substance was of course petroleum, and it was no new discovery. People in many countries had been aware of its existence, as it seeped out of the ground or bubbled up in streams, as it did at Oil Creek near Titusville, in Pennsylvania. Some of its properties had been known since ancient times. Bricks in the walls of Jericho, dating from perhaps 8000 BC, were bonded together with mortar made from bitumen, a form of petroleum. The Egyptians of the Old Kingdom had used bitumen for embalming the dead as early as the fifth millennium BC. The Chinese were probably the first to use petroleum for industrial purposes. There are accounts of workers drilling for brine in the time of Confucius, around 600 BC, and striking oil and gas. They apparently found little use for the oil, but ran the gas through bamboo pipes and used it to illuminate the salt mines. More recently, the crude oil which seeped from the ground around the Russian port of Baku on the Black Sea was burned for fuel in Czarist times.

The first man in history, however, to deliberately seek out this underground resource was an unemployed train conductor named Edwin Drake. He had been engaged in 1858 by a group of eastern investors to drill at Titusville for this new oil, which was beginning to find a market as an illuminant and a lubricant. Drake knew little about drilling, so he hired the local blacksmith, 'Uncle Billy' Smith, who had experience with artesian water bores.

Drake and Uncle Billy set up a simple drilling rig, and erected a wooden slab building over it for protection from rain and sun. At first they made little progress, as the water-logged sandy soil kept blocking their drill hole. Finally, they drove a large cast-iron pipe ten metres down to bedrock and began drilling into the rock through the pipe. On Sunday afternoon, 28 August 1859, with the drill out of the hole, 'Uncle Billy' looked down the pipe and saw oil floating just below the level of the hut floor.

Drake's well produced only about ten barrels of oil a day, and it did not make a fortune for its owners. But it was the very first commercially

Above left: Although petroleum oil from natural seepages had been known and used for centuries, especially for lighting, the first man to drill for it was Edwin Drake (left). His drill, inside the wooden hut, produced a commercial flow of oil at Titusville, Pennsylvania on August 28, 1859. *Above*: Oil wells were soon gushing in the quiet valleys of northern Pennsylvania, setting off yet another stampede of fortune-hunters. *Left, top*: The mushroom oil town of Pithole sprang into existence within a few months of a strike. *Left, below*: When the wells almost immediately began to run dry, everyone moved on, and within a year of its birth Pithole had vanished as suddenly as it had appeared. *The site of Pithole is now a Historic Park, and the world's first commercial oil well and its hut have been rebuilt at the Drake Well Memorial Park and Museum in Titusville.*

organised oil well in history. To mark its historic location, Drake's wooden hut and his drilling rig have been reconstructed on the spot at Titusville where it originally stood.

When news of Drake's oil strike spread, it aroused an instant response among those numerous individuals in the United States who were ever ready to go fortune-hunting. And so began yet another hectic American adventure, as oil seekers in their thousands rushed to Pennsylvania. Oil was found along many of the creeks and valleys, and soon there were crowded prospecting and drilling camps everywhere. One astonishing boom-town was called Pithole. It had what was probably the shortest and most hectic existence of any sizeable human settlement in history.

The first strike was made near Pithole Creek, not far from Oil Creek, in January 1865. It produced eight hundred barrels of oil a day. Almost immediately, an even bigger well came in. It looked like a whole new field. By May, five hundred building lots and twenty-two streets had been laid out above Pithole Creek, and rapidly began to fill up with buildings – some completed in four or five days. Within weeks an entire town had materialised. It had two banks, two churches, two telegraph offices, a newspaper, scores of hotels, a piped water supply and nearly sixteen thousand people. Pithole City throbbed with excitement, day and night. It was America's first oil boom-town.

Then, in August, while new fortune-seekers were still pouring into town, the biggest well stopped flowing. Others also began to dry up, and departures soon outnumbered arrivals. Disastrous fires swept through the wooden streets. By January 1866, within the space of a single year, Pithole was deserted once again. Today the site is a pleasant, grassy park, dotted with trees. A small information centre tells the story, with evocative photographs of the town that rose and vanished without trace.

By now, all across the United States, the search for oil was on in earnest, to satisfy the large and growing market for petroleum products for heating and lighting, lubrication, and industrial solvents. Oil was found in Virginia, Tennessee, Colorado, California, Ohio, Oklahoma and Texas. With each new discovery the gushers spouted higher and the underground oceans of 'black gold' seemed to grow larger. John D. Rockefeller came on the scene and gained effective control of about one quarter of the oil industry for his new company, Standard Oil, by providing the most efficient refineries. With the growth of the refining industry and its technology came the capacity to create new petroleum products. The role of oil in world history was about to be sensationally enhanced, as the lightest fraction of the distillation process came to realise its potential. It was called gasoline.

As the United States entered the last decades of the nineteenth century, its people seemed agreed upon a set of powerfully simplified objectives. The most highly prized of them was self-enrichment through business, and to achieve it rich and poor alike were prepared to work together.

By this time the Americans had enshrined a whole pantheon of heroes of material success. They were ruthlessly efficient, self-made

men of commerce, the creators of wealth not only for themselves but for others – the Rockefellers, Morgans, Carnegies, Vanderbilts. They were hated by some but envied by many and emulated by most. By contrast, in England, the country which began the Industrial Revolution, it was not until 1886 that the first businessman of the proliferating middle class was made a peer. There was still a reluctance to share status and privilege with the men of 'muck and brass'. By then, in the United States, there were four thousand millionaires, and where there was no 'old' money, 'new' money was just as good.

By the 1880s the Americans were poised to reshape not only their own history, but that of much of the Western world. There was capital by the billions, waiting for investment. There was a huge, single internal market a continent wide, hungry for things which would make work easier and life more comfortable. A swelling tide of talented, eager immigrants assured both the supply of labour and a continued dedication to the quest for material possessions and the good life. There was a spectacularly cheap and versatile new source of energy – petroleum. And all over the West there were seemingly boundless deposits of iron, copper, gold, silver and every other mineral that might be needed.

Armed with such human and material resources, the Americans embarked upon a vast application of technology in almost every field of human existence. To this they brought their irrepressible enthusiasm, their 'Yankee' ingenuity and their undeniable skills with their

In America's first oil boom, forests of derricks sprouted across the continent as the underground oceans of 'black gold' were tapped. This was Huntington Beach, California. *Huntington Beach is now a suburb of Los Angeles and still has scores of small pumps sucking away in streets and gardens.*

hands. The results oozed out all over the map of the world, to transform, almost beyond recognition, the way in which we live. And at the heart of that burst of innovation lay the underlying American will to democratise successful ideas. They had an unshakeable conviction that they were taking the mass of other Americans into partnership with them in the admirable enterprise of making money, and making life easier.

Among the most dedicated adherents to this peculiarly American ideal was a man who was born on a Michigan farm in 1863, the son of immigrant Irish stock. His pursuit of a single objective was responsible for one of the most significant freedoms ever conferred on the common man – the freedom of movement. His name was Henry Ford.

Ford was sixteen when he left the farm and went to the nearby city of Detroit to become an engineering apprentice. Already he had a consuming aim. While only thirteen he had seen a portable steam engine trundling along a country road under its own power. It had formed in his mind a vision that he would never lose – a self-propelled conveyance, or horseless carriage, at the disposal of everyone who wanted one.

For some years young Henry worked in various engineering jobs, and learned a lot about machines and steam engines. Then, in 1885, he was called upon to repair a new type of engine. It was the internal combustion engine patented by Dr Nicolaus Otto in Germany. This, Ford decided, was what he needed for his horseless carriage. The concept continued to grow in his head for the next ten years, as his career developed.

By 1894 Henry Ford was chief engineer with the Edison Illuminating Company, one of the three main electricity suppliers to the city of Detroit. He was thirty-one, married, with a son, Edsel. He had a secure future, and was a respected figure in his community. But more important to him than all of that was the knowledge that in America, as in Europe, the biggest challenge in engineering was to make a workable 'horseless carriage'.

Ford knew that in Europe, in particular, such vehicles were already in use. He may even have seen the Benz 'motor carriage' exhibited in Chicago in the summer of 1893. But expensive, hand-built machines like these were clearly destined for a limited market among the wealthy. Ford had other ideas, and in a small brick shed at the back of his house in Detroit he began to put them into effect.

Ford began by building a small engine to run on gasoline. It was derived in general design from Otto's engine, but was greatly simplified. Ford used one cylinder, made from a small gas pipe, and a piston and crank similar to those he had seen on steam engines. The handwheel from an old lathe served as a flywheel, and the spark plug was a piece of fibre with a wire pushed through it. Ford brought his engine into the kitchen on Christmas Eve, 1893, and mounted it on the sink, with the spark plug connected to the electric light overhead. With his wife Clara pouring gasoline into a spout, Ford spun the flywheel. The lights flickered, the engine coughed, then burst into noisy life. Henry Ford had bred the artificial horse for his horseless carriage.

One of the most far-reaching consequences of the invention of the internal combustion engine and the adoption of petroleum oil as fuel was the evolution of the automobile. *The world's first self-propelled motor was this Benz tricycle, 1885, now in the Deutches Museum, Munich.*

Over the next two years Ford slowly evolved a vehicle built around a two-cylinder gasoline engine. He had intended making it a two-wheeler to start with, but when this turned out to be impracticable he built what he called his 'quadricyle'. It was a large box on four bicycle wheels. The engine was inside the box and drove the rear wheels through a chain. Ford sat on top of the box in a saddle seat, and steered the front wheels with a tiller.

Very early on the morning of 4 June 1896, Ford was ready to try out his quadricyle. When he tried to roll it out of his workshop he found that it was too wide to fit through the door. He impatiently smashed away some bricks, and a few minutes later he was sputtering in stately fashion along Detroit's Grand River Avenue, accompanied by a friend on a bicycle.

It was to take Henry Ford another thirteen years, and a succession of steadily improving models, to bring out what became the most famous motor car ever built – the Model T. It was the fulfilment of his vision of cheap mobility for everyman. And although Ford did not invent either the motor car or the assembly line, it was his adaptations of both which created the industries, linked to the motor car, which have come to dominate the economies and measure the well-being of nations.

Henry Ford did not invent the automobile or the assembly line, but his adaptations of both to the objective of providing personal transport and freedom of movement for the common man created industries which have come to dominate the economies of entire nations. *Michael Charlton and the Model T Ford, perhaps the most famous automobile ever built.*

Ford's insight was to see that modern technology could only be applied successfully to the assembly of large and complicated artefacts such as a motor car if the process was broken down into many component parts, to be carried out by individuals. 'The man who puts in the bolt doesn't put on the nut, and the man who puts on the nut doesn't tighten it,' he said. In that crisp injunction, Ford summed up the assembly line and mass production.

With such division of labour the five thousand parts of the Model T Ford could be put together eight times faster than they had before. When Ford introduced the idea of moving the vehicle instead of the men, the assembly time was halved again. Output shot up, and within five years Ford was able to cut the price of the Model T by almost half. It sold for as little as $290. The men who built it could afford to buy it, whereas in most of previous history the productivity of labour had been so low that the benefits of leisure and luxury could be enjoyed only by the privileged.

In 1913, the year the moving assembly line was installed, the Ford plant was turning out one thousand cars a day. Production peaked in 1923, when two million cars were made. Altogether, Ford produced more than 16,500,000 Model Ts. It was the subject of more jokes, stories, controversy and affection than any other car in history. When the last model rolled off the line in 1927, its predecessors had rattled their way to every corner of the world, and many are still going.

Much has been written about Henry Ford's practical and organisational capacities but there was another, lesser-known reason for his success. What made the Model T possible, at a time in history when most vehicles were still being made of wood, by hand, was Ford's grasp of metallurgy and the manufacturing advantages of precision-made

210

metal parts. Ford was one of the first to appreciate and use the special properties of metals to be cast, forged, pressed or machined to any pattern, to tolerances which would make parts truly interchangeable, and thus capable of assembly by relatively unskilled workers.

Ford was also extremely knowledgeable about the behaviour of metals and their alloys. One essential factor in the success of the Model T was its combination of lightness with strength, which Ford achieved by the use of a highly advanced alloy – vanadium steel – in its axles, transmission and key engine parts. Since Ford had no formal training in metallurgy, it is not certain how he became familiar with such exotic materials. Ford himself later said that he had been watching a motor race in Florida when a French car had an accident. He picked up a broken piece of valve gear, and out of curiosity had it analysed for comparison with American valves. It was found to be made of an alloy of steel and vanadium, then a quite rare metal.

Of more significance, perhaps, was the appearance at Ford's plant in 1905 of an Englishman named J. Kent Smith. He was promoting British vanadium steel for its strength and resistance to shock and fatigue. Ford later established his own metallurgical laboratory to study such alloys, and found the manufacturing problems greatly eased by the fortuitous discovery of large deposits of vanadium in South America.

Towards the end of his life, Henry Ford became notorious for his intransigent attitude towards unions and labour problems, but in some of his innovative labour policies early in this century he was far ahead of contemporary views. In 1914 he rocked the industrial world by introducing the eight-hour day and an employee profit-sharing plan which included a basic daily wage of five dollars for all his workers – both men and women. This was double the going rate of pay, and while industry predicted bankruptcy the best workers in America flocked to Detroit and the Ford plant.

Ford's initiative, when it became widely adopted, put huge spending power in the hands of the workers, a factor which fuelled the growth of the American economy. And although in later years the assembly line was to become a metaphor for the loss of human individuality in the industrial society, it put material possessions within reach of millions for the first time, by making manufactured metal and plastic goods both cheap and widely available. Hand-made products of wood, leather or natural materials gave pride to the maker but could never have supplied the wants of the new urban populations, rapidly being bred out of their past self-sufficiency.

Apart from his own contribution to the 'American dream', Henry Ford was a remorseless collector of machines and gadgets from that enormously innovative period of American history. That era has been embalmed for us in the vast museum complex that Ford dedicated in 1929 at Dearborn on the outskirts of Detroit.

Henry Ford's astonishing tribute to American practical inventiveness comprises two parts, Greenfield Village and the Ford Museum. The former is a sprawling pastoral vision of early America, and includes many historic buildings, such as Edison's laboratory and the Wright brothers' bicycle shop, brought together from all parts of the United

States. The Ford Museum takes its theme from the inscription that Henry Ford fixed over the doorway of his first laboratory: 'Mankind passes from the old to the new over a human bridge formed by those who labor in the three principal arts – agriculture, manufacturing, and transportation.'

Here, under one all-encompassing roof, are the jigs, tools and dies which punched out the pattern of twentieth century life. In many cases they are the originals, the prototypes, of that great spawning of devices and machines designed to alleviate effort and burden which constituted the crucial objective of American culture towards the end of last century. Here is McCormick's harvester, Morse's telegraph, Edison's electric bulb, Bell's telephone, Eastman's camera and Ford's quadricyle.

Among the thousands of exhibits, which range in size up to a six hundred tonne steam locomotive, there are a handful which illustrate that one objective to which the Americans applied their greatest enthusiasm – the mechanisation of work in the home and the office, as well as in the factory, on the farm and down the mine. Here, for example, is the first Singer sewing machine, dated 1854; the Remington typewriter, 1874; the Bell wall telephone, 1891; the Thor electric washing machine, 1907; the electric iron, 1908; the electric toaster, 1909; and the Eureka vacuum cleaner, the Fearless dishwasher, and the Hughes electric stove, all 1910.

As the 19th century drew to a close the American genius for mass production, and the compulsion to popularise technical advances, loosed on the world a flood of labour-saving devices. *The first washing machine, introduced in 1907, and the prototypes of the vacuum cleaner (1910), the sewing machine (1854), the dishwasher (1910), the electric iron (1910), and the typewriter (1874), form part of the huge collection of technology in the Ford Museum at Dearborn, near Detroit.*

These labour-saving devices were not all invented in America; in fact most of the concepts originated in Europe. But in America they were first made practicable and workable, and manufactured on a scale and to a price that made them available to the world at large. It is to the Americans, and their belief that innovation and technology find fulfilment only by being shared, that whole classes of workers, and particularly women, owe their liberation from traditional drudgeries.

If the effect of the Ford Museum is overwhelming, the impression of the leafy streets and pre-industrial calm of Greenfield Village is at first deceptive. There is little indication in the modest nineteenth century facades of the houses of the momentous events that were set in motion by those who lived in them.

One elegant white-painted wooden house, with curtained windows, came from Dayton, Ohio. For forty years it was the home of a family whose two sons ran a bicycle shop. The brick-fronted shop, with a lean-to workshop behind it, stands beside the house with a solitary bicycle in the window. It was in this shop that Wilbur and Orville Wright not only made and sold bicycles but conceived, designed and built the first successful heavier-than-air, powered flying machine, which catapulted the world into the space age.

The workshop is just as it was at the turn of the century. On a bench lie a few ribs from an aircraft wing. There is an upright drilling machine, tools hanging on the wooden walls and in one corner a cut-down wooden propeller mounted on a shaft and driven by a belt from an electric motor. The Wright brothers used this device to blow air through a square wooden tube, standing next to the doorway. This was the prototype of the wind-tunnel, and it was a crucial tool in their most significant contribution to the achievement of powered flight – control of stability and direction in three dimensions.

The realisation of one of man's most elusive dreams – the freedom of flight – had its roots in the perseverance of men like Otto Lilienthal in Germany, who made hundreds of descents with the forerunner of the hang-glider before he was killed in 1896. *This diorama forms part of the historical display at the Air and Space Museum in Washington.*

The first powered, heavier-than-air machine to fly and manoeuvre in the atmosphere, the Wright brothers 'Flyer', opened the way to what the Americans regarded as their last frontier – space itself. *The world's first true flying machine now hangs in the Air and Space Museum in Washington, next to the first US space capsule.*

The heavier-than-air flying machine had its origins long before the Wright brothers, and in fact many of its theoretical specifications had been set out in England by Sir George Cayley, the 'father of the aero-plane', as early as 1799. Cayley envisaged the basic configuration which is now considered standard – a body or fuselage supporting one main wing, with horizontal and vertical control surfaces on the tail. In the 1890s, Lawrence Hargrave in Australia was close to heavier-than-air flight with his box-kite concept, but lacked a suitable power source.

What the Wright brothers brought to the challenge was an intensive programme of research, engineering and testing, in typical American fashion. From their own experiences with gliders, and from the published work of others like Otto Lilienthal in Germany, who invented hang-gliding and was eventually killed by it, the Wright brothers knew that the major difficulty was to achieve stability through the correct combination of control surfaces. There was little or no data to draw upon, and much of it was misleading.

To understand these problems, the Wright brothers conceived the idea of making tests in a wind-tunnel. They did not actually invent this device, but they used it more specifically than anyone before them to study the behaviour of various model wings and control surfaces in simulated flight. Only then did they proceed to construct their 'Flyer'.

The Wright brothers came to appreciate the mysteries and subtleties of powered flight so much more thoroughly than other experimenters that after their first historic hops on 17 December 1903 – the longest being fifty-nine seconds – it was four years before anyone else was able to stay aloft as long as a minute. Even as late as 1908, when the Wright brothers flew publicly, circling and turning for more than an hour, no one else knew how to execute proper turns or make efficient propellers.

That first halting flight at Kittyhawk was one of the greater lurches of history. It changed, after all, the conduct of both peace and war. It should be considered therefore not a surprise but almost inevitable that what the Wright brothers were the first to pioneer, and which was so widely taken up by others, would point the way to the place which knows no boundaries, and which the Americans came to regard as their last frontier – space itself.

But at the opening of the twentieth century all that still lay ahead. Already, however, it was clear that in freeing mankind from the bonds of gravity a fundamental nexus had been broken. The challenge of flight would stretch the applications of metallurgy to new and unknown limits. In all of history, to this moment, metals from the earth had been Earth-bound. To make them fly would take a new kind of specialist – the scientist.

CHAPTER 10
From alchemy to the atom

At least ten thousand years have passed since man first began to use metals in their natural state. And perhaps eight thousand years have elapsed since he first began to smelt metals out of the rocks. But using the metallic elements was one thing; understanding what distinguished one from the other, and comprehending why they behaved as they did – all this had to await the Age of Science.

True scientific understanding of the universe and the elements of which it is made is surprisingly recent. It is still less than two hundred years since John Dalton, an English schoolmaster, worked out a mathematical relationship for the atoms of matter, and laid the foundations of modern physics and chemistry. Of course, there were theories long before Dalton.

The earliest recorded speculation on the nature of matter is by Thales, the 'father of Greek philosophy', who lived in the seventh century BC. He suggested that water is the primordial element from which all matter is derived. In the fifth century BC, Anaxagoras began to talk about infinitely divisible 'seeds' which combine to form solid bodies, while Democritus put forward the first 'atomic' theory, holding that matter is created by the chance collision and aggregation of immeasurable, invisible and homogeneous particles. A century later Aristotle suggested that the fundamental elements are built up from four properties – heat, cold, dampness and dryness – combined in various ways. Heat and dryness produce fire; heat and dampness, air; cold and dryness, earth; cold and dampness, water.

By the Middle Ages many strange and wonderful notions about the origin of the elements, and the source of metals in particular, had been erected on Aristotle's theory. There was great interest in his suggestion that if the relative proportions of the properties of an element are altered, one substance may be changed into another. It was this idea which fascinated those hopeful speculators, the alchemists.

The secretive world the alchemists inhabited is vividly re-created at

Left: From the Greek philosophers onwards, man has been fascinated and puzzled by the nature of matter and of the atoms which make up the universe. *Mineral specimen from the Geological Museum, Sydney. Above*: The alchemists in their shadowy dungeons were among the first to study the nature of matter, as they sought to transmute base metals into gold. *The Apotheken Museum in Heidelberg Castle in Germany, although devoted to the early apothecaries, conveys an eerie sense of the world of the alchemists.*

the Apotheken Museum in Heidelberg Castle, in Germany. The enormous pile of the castle, begun in the thirteenth century, dominates the ancient university city from the heights overlooking the Neckar River, near its junction with the Rhine. With its sheer stone walls and partly-destroyed towers and turrets it embodies a brooding medieval presence. This feeling is conveyed even more strongly by the museum in the castle devoted to the work of the ancient apothecaries. The narrow stone rooms are lined with flasks and bottles of herbal remedies, rare earths and mysterious potions – the pharmacopoeia of the past, yet not so far removed from the drug-lined shelves of a modern pharmacy.

The echoing dungeons of the museum, with their furnaces, oddly-shaped glass retorts and strange implements, summon up the dark practices of the alchemists, as they sought among the flames and crucibles an explanation of their universe. The chief obsession of the alchemists, and the subject of their most devoted investigations, was the transmutation of one metal into another. In particular, they pursued the elusive secret of turning base metals into gold.

The alchemists held a central belief, that in the molten fires beneath the Earth's crust new stores of metals were constantly being created by the direct influence of the planets (in which category they included the Sun). According to this view, each planet was responsible for developing or bringing into existence a specific metal, and upon that element impressed its own characteristics. The Sun, for example, influenced the formation and nature of gold. The Moon was responsible for silver,

Mars for iron, Venus for copper, Saturn for lead. Once the metals were created within the earth they became involved in a continual process of transformation from a lower to a higher state. Lead thus became silver, and silver finally became gold. That ultimate transmutation was what the alchemists sought to emulate, using fire to re-create the furnace in the interior of the planet.

With the invention of the printing press, the writings of the alchemists multiplied at an astonishing rate, and their teachings made an impression which survived for a surprising length of time. A compelling motive in the despatch of the Spanish conquistadors to Mexico and South America in the sixteenth century was the alchemists' theory that gold would be found most abundantly in those regions where the Sun's rays fell upon the Earth with the greatest intensity.

But the fantasies and concoctions of the alchemists in the end came to nothing. There was no gold at the end of their labours in sulphurous dungeons, and their efforts remain among the less useful additions to our store of knowledge. Their dreams were not to be realised until our own times, when the nuclear physicists began to investigate the radio-active metals. Using the neutron to unlock the greatest doors, which had remained firmly shut in the face of alchemy, they gained the awesome knowledge of how to transmute elements not into gold but into something much more potent – energy. First, however, there were profound mysteries about the nature of the Earth itself to be answered.

The question of how the metals and other materials of the planet came to be formed and distributed in such apparently random fashion continued to puzzle and perplex philosophers until well into the seventeenth century. Only then did the obscurantist mists begin to clear, with the new enlightenment of the Renaissance, and the first real insight into the nature of the Earth and its rocks and metals begin to form.

A fundamental contribution to the new way of looking at the world around us was made in Edinburgh. By the eighteenth century this grey granite city had become one of the major centres of thought and learning in Europe, with its rigorous education system, its Royal observatory and medical school and, at the centre of its academic life, its great university. In this environment of progress and scientific enquiry there came together a group of brilliant and innovative Scots, who were all pupils of a remarkable professor of chemistry, Andrew Plummer.

Little is known about Plummer, but he must have been an outstanding teacher to have so successfully developed the curiosities and talents of such a group. One of them was Dr Joseph Black, whose theory of the latent energy of heat was a crucial inspiration to James Watt in his development of the first true steam engine. Another was James Hutton, whose theories about the Earth, when first put forward, were labelled as heresies. In the end they prevailed, and gave us our first understanding of the slow but inexorable cycle of continent-building. Hutton laid the foundations for a new branch of science, of which he is now acknowledged to be the father – the science of geology.

James Hutton was an eighteenth century all-rounder, typical of those times. Of independent means, he was in turn farmer, lawyer, physician

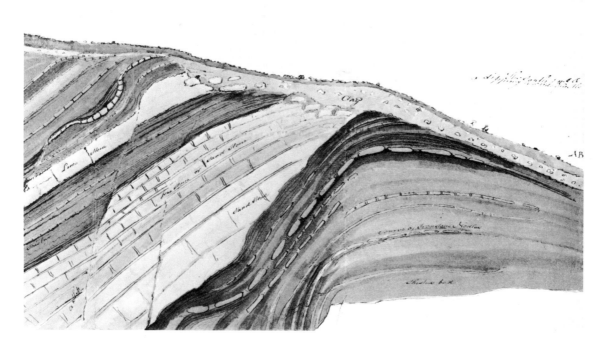

and chemist. Above all, he was a curious and observant man. Living as he did in Edinburgh, his gaze was forever attracted to Arthur's Seat, the blunt, flat-topped rock formation that dominates the city. Hutton's interest was aroused by the way the rocks were arranged in quite distinct layers, like a cake. His insight led him to seek, in this pattern, an explanation of the past.

Hutton spent many years studying and recording rock formations in various parts of the Scottish highlands, and gradually constructed the coherent architecture of a new concept of how the landscape itself is shaped and modified. In 1785 he published *Theory of the Earth*, illustrated with drawings of the various kinds of rocks he had observed. In this work he set out to interpret what he called 'the beautiful machine' which constantly creates, wears down and renews the surface of the planet.

What Hutton proceeded to do was to give the first unified theory of the slow cycle in which 'new' rock wells up from within the earth to form the mountains and the continents, only to be worn down again by erosion and deposited in layers of sediment on the ocean floor. This ancient material, when uplifted by movements of the Earth's crust, forms the sedimentary rocks which we see all around us. It was the end result of this cycle which Hutton first recognised in Arthur's Seat.

Hutton's theory clashed directly with those of the 'Neptunist' school of thought, led by the German Abraham Werner, which held that the whole planet had once been covered by a deep ocean, and that the rocks which now form dry land were deposited from that universal ocean by chemical precipitation. Hutton's views also conflicted inevitably with existing theories of Creation. Hutton's comment in his book that in nature he found 'no vestige of a beginning, no prospect of an

James Hutton's book, *Theory of the Earth*, illustrated with drawings of rock formations around Edinburgh, at first aroused controversy, but in the end brought him enduring fame as the 'father of geology'. *This illustration is from a set of prints from Hutton's book, supplied by the Department of Geology at Edinburgh University, where the originals are kept.*

end' was far too heretical for a society which had yet to be dissuaded from Bishop Ussher's conviction that the world had come into existence in 4004 BC.

By the beginning of the nineteenth century, however, and largely because of Hutton, rocks everywhere were coming under more scientific scrutiny. One informed observation in Provence was to involve this historic region of France in a chain of events which yielded to the world a new and uniquely endowed metal.

The striking geological formation called Les Alpilles, near Avignon, is a series of jumbled limestone ridges and crags, created long ago by the crumpling of the landscape between the Pyrenees and the Alps. On one narrow, sheer-sided plateau rising above the plain stands a ruined fortress. A white-walled, red-roofed village clings to the slope leading up to the citadel. Called Les Baux, this rocky stronghold was one of the great redoubts of medieval Europe. It dates from a time, a thousand years ago when perspectives were limited, and those who held the heights dwarfed all those who lived on the plains.

A chance observation in 1821 of a band of red earth in the white limestone formations around the village of Les Baux in Provence led to the discovery of the first large deposit of aluminium oxide. This ore was named after the village, and has been known as bauxite ever since. *Bauxite is still being mined at Les Baux, although the open-cut is now restricted to preserve the character of the area and all extraction takes place underground.*

Les Baux takes its name from a powerful family which in the tenth century took possession of the highest point of the plateau, with its traces of Celtic and Roman occupation. They dug and carved and pierced the white limestone, making a huge, extraordinary fastness, half cave, half palace. For more than five hundred years the lords of Les Baux – 'a race of eaglets, never in bondage' – dominated not only their eyrie but the surrounding countryside. They were 'burglars of monasteries, who brought murder and fire, and passed through bastions and castles like an iron rake'.

220

The power of the lords of Les Baux was finally broken in the fourteenth century, and others moved on to the heights to build houses and churches. In the fifteenth century Provence was united with France, and soon afterwards the French king created a line of barons of Les Baux. In the seventeenth century, however, the last baron joined an unsuccessful rebellion against Louis XIII, and in retaliation Cardinal Richelieu had the great citadel of Les Baux demolished with gunpowder and hammers.

For two more centuries the ruined citadel and its attendant village lay in the sun of Provence, visited only by poets attuned to the echoes of the troubadours who once strolled the battlements with the ladies of Les Baux. The only change occurred inside the towering ridge, where stonecutters sliced out blocks of limestone for buildings in Avignon, leaving vast, echoing caverns with lofty stone doorways, like Pharaonic temples. Then, early in the nineteenth century, there came to Les Baux an observer who looked at the landscape with different eyes.

It is well to be properly sceptical about legends of accidental discovery, particularly in science. Serendipity is more often the reward of painstaking advances from established platforms of knowledge. But the allies of such advances are invariably curiosity, observation and a willingness to get off the beaten track. Such a byway through the tumbled formations near Les Baux was taken in 1821 by a young chemist named Berthier. He was on a walking holiday in Provence, and had come to Les Baux because he was an amateur geologist.

Berthier was exploring the limestone ridges when he saw something dramatically conspicuous against the dazzling white of the landscape. It was a vivid red horizontal band in the face of a cliff. Berthier examined it closely, because it resembled nothing he had ever seen described in books about rocks. There is no record of whether he recognised it for what it was – a remarkably rich deposit of a mineral which had been both provoking and intriguing scientists in Europe for centuries.

However, the name that Berthier gave his find created an enduring link between this historic region of France and what was to become one of the most important elements in the modern world. Berthier named the unusual red material after Les Baux, and bauxite is now the universal name for the ore which yields the metal which speaks most eloquently of our times – aluminium.

Aluminium can truly be described as a modern metal – and not only because its special properties of lightness, strength and resistance to corrosion have made it indispensable in the age of flight. It has been available on an industrial scale for less than one hundred years. This is remarkable, considering that it is the third most common element in the Earth's crust, after oxygen and silicon. Being present in the proportion of 8.3 per cent, aluminium is easily the most abundant metal, well ahead of iron at 5.0 per cent. It occurs almost everywhere, especially in clays, and not just in commercially available red bauxite deposits. And yet it was not until the nineteenth century that European scientists were able to isolate aluminium in metallic form.

The problem with aluminium is that it never exists in a native or pure metallic state, but is invariably oxidised. The atoms of aluminium are bound to oxygen atoms (and sometimes to silicon as well) with chemical links so strong that no normal method of reduction or smelting can separate them. Thus during the first few millennia of the Age of Metals, as man learned how to use copper, iron, gold, silver, lead and zinc, the most plentiful metal of all remained unknown.

But if the intractability of aluminium to smelting delayed its use in metallic form until modern times, it did not prevent early man from making use of aluminium compounds. Aluminium is an important element in many attractive and precious gem stones – rubies, sapphires, emeralds, aquamarines, turquoise, lapis lazuli, tourmaline, topaz and jade. Another common aluminium compound is alum, whose astringent properties were known to the Egyptians and Babylonians; as early as the third millennium BC it was used to control bleeding and to fix dyes. By Greek and Roman times this substance had been given the name 'alumen'.

The question of when metallic aluminium first came into use cannot yet be answered with certainty. The Roman historian Pliny the Elder (AD 23–79) included in his monumental 'Natural History' the claim that 'a metal whiter than iron, that could be worked like silver', was obtainable from clay, but added that the Emperor Tiberius (AD 14–37) had ordered the discoverer to be beheaded 'so that his gold would not be depreciated by the new metal'. There is no indication of how this interesting metal had been extracted.

Even before this, according to recent discoveries, the Chinese may have been reducing metallic aluminium from its ores. The question was raised by the weapons recovered from the famous terracotta army of the Emperor Ch'in, discovered near Xi'an in 1974. No Western scientific studies of the weapons were immediately possible, but the Chinese archaeological authority in charge of the work published metallurgical analyses of a sword and several arrow heads. They are described as containing, in a complex copper alloy, a measurable amount of aluminium. There is no known way that aluminium could have been reduced accidentally during the smelting of other metals.

What can be said with certainty is that by the Middle Ages the alchemists and philosophers were arguing about the nature of a mysterious substance which seemed to exist in alum and certain earths, and which came to be called alumina. In the thirteenth century the English natural philosopher Roger Bacon suggested that the basis of alumina might be a metal. No further progress was made until the eighteenth century. Then, in 1782, the great French chemist Lavoisier wrote that he suspected alumina to be 'the oxide of a metal with such a strong affinity for oxygen that it is not reducible to metal by carbon or any known methods of reduction'.

The first close approach to the production of metallic aluminium was made by Sir Humphrey Davy, the eminent English scientist. Davy had been experimenting with the latest invention by the Italian, Alessandro Volta – the electric storage battery. He had successfully broken down various chemical compounds by passing electricity through them in the process called electrolysis, and applied this treatment to a mixture containing alumina. When this did not work, he tried a mixture of iron

Left: Aluminium is smelted by a continuous process in pots containing chemicals, kept molten by large currents of electricity passing between carbon electrodes. The aluminium metal collects in the bottom of the pots and is tapped off at intervals. *Carbon electrode being changed at Pechiney smelter in the French Alps.*
Above: When first produced, aluminium was more precious than gold, and reserved for special objects. The aluminium for this ceremonial helmet, made for King Frederick VII of Denmark, was more expensive than the gold which decorates it. *The helmet is on display in Copenhagen.*

and alumina, and applied the electricity in the form of an arc discharge. This produced something which Davy had never seen before.

In 1809 Davy reported to the Royal Institution that he had obtained a small globule of a 'white alloy, harder than iron', and added that 'if I had been fortunate enough to isolate the metal which I was striving to obtain I would have given it the name alumium'. Davy later modified his choice of name slightly to aluminum and finally to aluminium. (In 1925 the American Chemical Society decided that the correct spelling should be aluminum, and this is why North American usage differs from the rest of the world.)

Others continued the pursuit of the elusive metal, and in 1825 Hans Oersted in Denmark obtained a greyish powder, which enabled him to describe aluminium as having 'the colour and lustre of tin'. But it was not until 1845 that the German chemist Friedrich Wohler succeeded in producing the very first globules of aluminium, of sufficient purity for him to describe the malleability and lightness of the silvery metal. Wohler's method was a chemical one, involving a reaction between aluminium chloride and potassium vapour. A similar chemical process was used by the first man to produce aluminium in any quantity – the French chemist Henri Etienne Sainte-Claire Devillc.

Deville had been supported by the Emperor Napoleon III in his work on aluminium, and on 6 February 1854 he performed a famous demonstration at the Academie des Sciences in Paris. Using a refined version of Wohler's method, Deville produced not only isolated globules but a small ingot of pure aluminium. Soon afterwards he demonstrated a method of reducing aluminium by electrolysis, along the lines pioneered by Sir Humphrey Davy, but abandoned it because of the high cost of storage batteries, which were then the only source of electricity.

Deville confirmed his claim as the 'father of aluminium' by steadily increasing his production of it, until in 1857 he was turning out fifty kilograms a day. But aluminium remained a rare and comparatively expensive metal, and was much admired for its lustrous, silver appearance. A bar of it was exhibited at the Paris Exposition of 1855 – next to the English Crown Jewels. Napoleon III was highly taken with the new metal, and asked Deville to make him a breastplate, a service of forks and spoons for State banquets and a rattle for his infant son. The Emperor was also one of the first to put to use the lightness of aluminium. He had the bronze eagles on the standards of the French army replaced by ones cast in aluminium, to lighten the load of the standard bearers.

Aluminium production received its first significant boost in 1858, when a mining engineer in Marseille named Meissonier sent Deville a sample of iron ore which he was finding difficult to smelt. Deville analysed the red material and found the 'iron ore' to be a high grade aluminium compound. The sample turned out to have come from the deposit of bauxite which Berthier had discovered and named nearly forty years before. The exposed seam was found to extend for many kilometres under the limestone hills (and it is still being mined commercially). With ample supplies of bauxite from Les Baux assured, Deville went into commercial production with a foundry at Salindres, in Paris.

By using cheaper chemicals and improving the reduction process, Deville and others steadily reduced the cost of aluminium. From three thousand gold francs per kilogram in 1854 the price fell to one thousand francs in 1855, and to three hundred francs by the end of 1856. By the 1880s production in various European countries had risen to some thousands of kilograms a year, and the price had stabilised at about forty francs per kilogram. Aluminium was still comparatively expensive but its remarkable properties were becoming widely known, and intensive research continued into cheaper ways of producing the silvery metal.

The answer, when it came, involved one of those extraordinary coincidences which sometimes occur in scientific research. Two young men born in the same year, 1863, on two different continents, worked out precisely the same innovative method of producing aluminium, quite unknown to one another. Both applied for their patent in the same year, 1886 – and both died in the same year, 1914.

The two were Paul Heroult in France and Charles Hall in the United States. At the age of twenty-two, the newly-graduated students tackled the challenge of finding a cheaper way of making aluminium. Neither was a professional chemist, nor were they employed by large metallurgical companies. Heroult did his research in a small room in the corner of a Paris tannery. Hall was the son of a Congregationalist minister in Ohio, and set up a makeshift laboratory in a woodshed behind the parsonage.

Both chose to turn away from the Deville chemical process then in general use and go back to the electrolytic method first tried by Sir Humphrey Davy, eighty years before. Both hit on the same solution which was to dissolve the alumina in a bath of molten cryolite – an unusual mineral consisting of fluorine, sodium and aluminium – and pass massive currents of electricity through it. The energy provided by the electricity kept the mixture molten, and also reduced the aluminium from its ore. The aluminium sank to the bottom of the bath, where it could be tapped off and cast into ingots of pure metal. The process was continuous as long as fresh alumina was added to the cryolite bath.

The Hall–Heroult process is the basis of today's world-wide production of nearly twenty million tonnes of aluminium a year. This makes aluminium the second most important metal after iron and its alloys. Its career really began with the decision by Hall and Heroult to resuscitate the electrolytic method of smelting aluminium ore, which had previously proved uneconomic. What made them take such a decision? It was the appearance on the scene of something which was not available to Sir Humphrey Davy – a cheap and sustained supply of electricity. By another of those strange coincidences, this had become possible through a discovery made in London in 1821, the very year that Berthier had stumbled across the first known commercial deposit of aluminium ore.

At the beginning of the nineteenth century the centre of scientific thought and experimentation in England was the newly-formed Royal Institution, off Piccadilly. In the century to come the names of those who spoke from the dais in the famous lecture theatre, with its semi-circular rows of seats rising steeply on three sides, would read like a roll-call of the rise of science. But among the many famous and often historic discourses and demonstrations delivered there, few had the far-reaching consequences of four lectures on chemistry given early in 1812 by Sir Humphrey Davy, then President of the Institution.

Davy was not only an outstanding scientist but a brilliant and hand-some lecturer of great charm and vitality, and in the lecture theatre of the Royal Institution he had created a large and enthusiastic audience for science, drawn from all sections of London society. Among these well-dressed people on the crowded benches in March and April, 1812, was a young man of twenty-one. He was an apprentice bookbinder, the son of a blacksmith. His name was Michael Faraday.

Faraday had left school at thirteen, painfully aware that he was not well-read and could spell only poorly. But he found a never-ending source of education in the books that he was given to bind. He read them all omnivorously, for he was dedicated to self-improvement. Sir Humphrey Davy, not unnaturally, was Faraday's hero. But had it not been for a thoughtful gesture by one of his master's customers who gave the young man tickets for Davy's lectures on chemistry, it seems highly unlikely that Michael Faraday, bookbinder's apprentice, would ever have become anything other than Michael Faraday, bookbinder. On the night of the first lecture Faraday took his place in the theatre, high up behind the clock, his notebook open, eager to learn what he could.

Davy's first lecture converted young Faraday into an ardent and passionate chemist. He finally wrote down all Davy's lectures in careful notes, bound them with particular care, and sent them to the great man. The President of the Royal Institution could hardly have failed to be impressed. In fact Davy was so impressed that he gave Faraday a job as assistant in the laboratory of the Institution at twenty-five shillings a week, and a tiny room at the top of the building to live in. The notes that Faraday took and bound, their pages now yellowing between the brown leather covers, are among the most prized possessions of the Royal Institution, where the tradition of Friday night discourses and lecture series continues, exactly as in Davy's day.

From that relationship between the famous scientist and the son of the blacksmith came momentous advances. Michael Faraday embarked upon a career of discovery in chemistry and metallurgy unparalleled in the history of experimental science. In one field, in particular, he showed the way to the practical application of the most useful form of energy in the modern world – electricity.

At this time there was a widespread fascination with the newly discovered, invisible force called electricity, which could make sparks jump through the air and give people nerve-tingling shocks. There was also curiosity about another mysterious force, first demonstrated by the Chinese compass needle, which had the power to attract iron. But just how closely these two forces of electricity and magnetism were related it was left to Faraday to demonstrate.

When electricity first came to public attention early in the 19th century it was little understood, but its mysterious capacity to make sparks fly and give people shocks made it the subject of elaborate parlour games. The 'Electrifying Kiss' was delivered by someone holding an electrically-charged brass frame. *This model is in the Smithsonian Institution in Washington.*

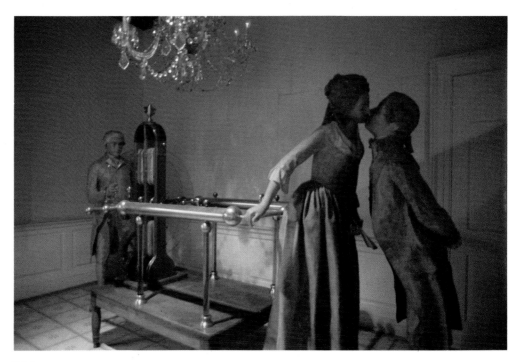

In 1821, in the downstairs laboratory of the Royal Institution, Faraday set up his historic experiment. On a bench he placed one of the early storage batteries, and near it two wine glasses filled with mercury, side by side. From a frame above the glasses were suspended two straight, stiff pieces of iron wire, which just touched the surface of the mercury in the glasses. One piece of wire, hanging loosely on a hook, was free to move around a small bar magnet fixed upright in the centre of the glass of mercury. The other wire was fixed with its tip in the mercury, but in the glass there was a bar magnet which was free to move around the tip of the wire. Conducting wires ran from the battery to the apparatus, making two circuits running through the hanging wires and the mercury.

When Faraday completed the first circuit by touching the connecting wire to the battery terminal, the loose hanging wire began to rotate around the fixed magnet in the glass of mercury. When he disconnected that circuit and connected the other one, the loose magnet in the second glass of mercury began to rotate around the fixed wire.

Faraday had discovered a completely new application of electrical energy, which produced rotary motion. It was the basis of all the electric motors which today drive a myriad modern devices, from the food mixer to the starter on a car, from the hi-fi turntable to the Japanese 'bullet train'.

The principle of electromagnetic rotations, as Faraday called it, depends upon the reaction between a magnetic field and a flowing electric current. It was a profoundly important demonstration that Faraday had devised, but he did not leave it there. For the next ten years he thought about the relationship between electricity and magnetism. He had a great feeling for symmetry in nature, and it seemed to him that if electricity and magnetism could produce movement, then movement and magnetism ought to produce electricity. In 1831 he showed this to be true, in an even more significant demonstration.

Faraday made his great discovery with apparatus of elegant simplicity – a bar magnet, and a coil of iron wire connected to a galvanometer, a device which indicates the flow of an electric current by the movement of a needle. When Faraday took the bar magnet and moved it backwards and forwards through the coil, there was a corresponding deflection of the needle of the galvanometer. Movement plus magnetism was producing an electric current in the coil. Faraday then found that he could produce the same effect by rotating a copper disc in a magnetic field.

The principle of electromagnetic induction was the second of Faraday's great contributions to our understanding and application of electrical energy. It is the basis of all the generators which light the cities and power the industries of the modern world. It was Faraday's demonstration which led to the invention by Siemens in Germany and Gramme in France of the 'dynamic generator', or dynamo. This made possible for the first time the production of electricity in sufficient quantities to develop the Hall–Heroult process for the smelting of aluminium.

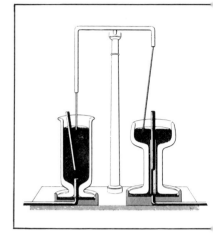

A contemporary drawing of the apparatus used by Faraday in his famous demonstration of the principle of electromagnetic rotations. In the right container of mercury there is a fixed, upright magnet, and the hanging wire is free to rotate around it. In the left container the hanging wire is fixed, and the magnet is free to move. When Faraday passed a current through the right wire into the mercury, the force operating between the flowing current and the magnetic field caused the wire to rotate around the magnet. In the left container, the current caused the magnet to rotate around the fixed wire. *This apparatus, which established the principle upon which the electric motor is based, is kept in Faraday's laboratory in the basement of the Royal Institution in London.*

Despite Michael Faraday's lack of early formal education, his great mental exertions and meticulous experimentation over more than forty years produced advances in chemistry, metallurgy and physics which have today put people everywhere in debt to the blacksmith's son. He opened a large window, letting in a flood of light which revealed new perspectives to all who followed him. His practical discoveries were the basis for so many of the everyday necessities that we now take for granted. The marble statue of Faraday beside the main staircase in the Royal Institution conveys something of the kindly, modest character of the first great visionary of the Electrical Age.[54]

But if Faraday was the first to demonstrate that electricity could be made to flow at will, the man who gave this new form of energy to the world in its most brilliant form was Thomas Edison.

It is easy today, in the age of electronics, to lose sight of the enormous contribution that Edison made to the easement of human existence. One reminder of it is his laboratory, which has been preserved intact with all its contents as part of the vast collection of American technology assembled by Henry Ford in Dearborn, near Detroit. The two-storeyed wooden building is crammed with equipment and apparatus which give some slight idea of the inquisitive vitality of one of the key figures of the heroic age of invention.

As they made a business of almost everything else, the Americans, in the latter half of the nineteenth century, made a business out of invention. In that landscape of enterprise, Thomas Edison was an Everest among foothills. In a career of invention which lasted for some sixty years, he took out more than a thousand patents. And in what he chose to do and how he brought his prolific intelligence to bear, Edison was the epitome of the democratic ideal in America. He had laid down social and political principles which guided his choices and which determined their priority. For Edison, the practical and the useful were paramount. The test of those qualities was what the market would accept. He made all his inventions conform to that command.

Perhaps Edison's best known invention was the speaking phonograph. Through the life-like preservation of the human voice, he offered resurrection after death. Here is one case where invention was not based on previous accomplishment. When Edison applied for a patent in December, 1877, nothing even remotely resembling his machine could be found in the by then vast records of the United States Patent Office. The idea of using a needle to inscribe grooves on a foil-coated cylinder, in order to reproduce the vibrations of sound waves by playing the needle back through the grooves, seems to have sprung fully formed into Edison's imagination.

However, in the realisation of his greatest feat of creative, applied intelligence, Edison freely admitted that inspiration had come to him in a straight line from Michael Faraday. When Edison was twenty-one he had read Faraday's account of his experiments with electricity, and it was in this field that Edison's own practical achievements were to shine so brightly.

Thomas Edison is remembered everywhere for inventing the electric light bulb, and certainly this was a technical triumph. It was the result

of an arduous research effort in which Edison tried literally thousands of materials – even the hair from the beard of one of his assistants – before he found a filament which would glow brightly when a current was passed through it, without burning out too quickly. The incandescent lamp was, however, just the beginning of Edison's most extraordinary burst of creative effort, which over a period of just eighteen months saw him devise and put into action the complete electric lighting system that we know today.

Edison had seen the potential of electric lighting, which had been demonstrated in the form of dazzling arc lights used to illuminate public spaces. But again the American mind leaped at the opportunity to democratise and apply an advance in technology. 'I saw that what had been done had never been made practically useful,' Edison wrote later. 'The intense light had not been subdivided so that it could be brought into ordinary houses . . .'

Edison saw that the answer lay in the conductivity of copper wires, which could be used to distribute the new form of energy to individual houses, just as pipes were used to distribute water or gas. The problem was that none of the components of such a system existed, and none of the theoretical calculations had ever been made. Edison himself has set down the magnitude of the challenge:

'Upon my taking up the electric light problem in 1878, my concept was a *complete system* for the distribution of electric light in small units in the same general manner as gas. And now the key-stone was provided [the incandescent lamp], it became necessary to prepare the other part of the structure (including the distribution of electric current for heat and power

Above: The dedication to research and the tremendous application of Thomas Edison is captured in this famous photograph, taken after he had spent several days and nights of virtually unbroken work to perfect his phonograph. *Left*: Another of Edison's great practical achievements, the electric bulb. *This replica of the original bulb was made by Edison himself in 1929, at the time his laboratory was moved to the Ford Museum at Dearborn, near Detroit.*

also). Some idea of the task may be gained from a perusal of the following partial program which confronted me:

'First – To conceive a broad and fundamentally correct method of distributing the current . . . so that in any given city area the lights could be fed with electricity from several directions, thus eliminating any interruptions due to disturbance on any particular section.

'Second – To devise an electric lamp that would give about the same amount of light as a gas jet, which custom had proven to be a suitable and useful unit . . . Each lamp must be independent of every other lamp . . . and remain capable of burning at full incandescence and candle-power a great length of time.

'Third – To devise means whereby the amount of electrical energy furnished to each and every customer could be determined, as in the case of gas, and so that this could be done cheaply and reliably by a meter at the customer's premises.

'Fourth – To elaborate a system or network of conductors capable of being placed underground or overhead, which would allow of being tapped at any intervals . . . With these conductors and pipes must also be furnished manholes, junction boxes, connections, and a host of varied paraphernalia, insuring perfect general distribution.

'Fifth – To devise means for maintaining at all points in an extended area of distribution a practically even pressure of current, so that all the lamps, wherever located, near or far away from the central station, should give an equal light at all times, independent of the number that might be turned on . . .

'Sixth – To design efficient dynamos, such not being in existence at the time, that would convert economically the steam-power of high-speed engines into electrical energy [and] means for regulating, equalising their loads, and adjusting the number of dynamos to be used according to the fluctuating demands on the central station . . .

'Seventh – To invent safety devices that would prevent the current from becoming excessive upon any conductors . . . also to invent switches for turning the current on and off; lampholders, sockets, fixtures, and the like . . . also interior circuits that were to carry current to chandeliers and fixtures in buildings.

'Eighth – To design commercially efficient motors to operate elevators, printing presses, lathes, fans, blowers, etc, etc . . . Motors of this kind were unknown when I formulated my plans.'[55]

Edison describes the months of feverish activity at his laboratory and workshops – 'we worked incessantly, regardless of day, night, Sunday or holiday' – as he and his team devised and made all the things he mentioned. 'Nowhere in the world', he wrote, 'could we purchase these parts . . . the only relevant item at this time was copper wire, and even that was not properly insulated.'

By Christmas, 1879, Edison was ready to demonstrate a complete electric lighting system. The Pennsylvania Railroad ran special trains from New York to Edison's laboratory at Menlo Park in New Jersey, bringing more than 3000 people to see the new light of the world. From the dynamo plant, underground conductors carried the current to Edison's laboratory, his office, twenty street lights and, most important of all, to Sarah Jordan's boarding house, where some of Edison's employees lived. It was the first domestic dwelling in the world to be lighted by electricity; and it is still standing with its original switches, exposed wires and globes in the grounds of the Ford Museum in Dearborn.

As first one and then another of the windows of the white wooden house lit up in the December dusk, a new prospect for millions gleamed through the darkness. Even disregarding the part played by electricity in every aspect of industry, commerce, and communications in the modern world, the domestic electric lighting system is one of the greatest single benefits that technology has brought to the daily life of people everywhere.

Within three years, Edison fired up America's first city power station, in New York. The Edison Electric Illuminating Company went into business on the night of 3 September 1982. There were eighty-five customers wired up to the system, and just four hundred electric lights to switch on. But what a beginning it turned out to be...

Edison's comprehensive achievement had many repercussions, but one of the most far-reaching was the new role that it created for copper, one of man's oldest servants among the metals. The properties of copper – the ease with which it can be drawn out into wire, its high electrical conductivity, and resistance to corrosion – made it ideal for circuits and switches. Henceforth, more than half the world's production of copper would be used in the service of electricity. In fact, so much copper has gone into the world's electrical systems in the past hundred years that it may take as long again for the new transmission materials, such as optical fibres and aluminium, to make much of an impression.

In that final quarter of the nineteenth century, in which the contributions of men like Thomas Edison, Henry Ford, Samuel Morse and Alexander Graham Bell came tumbling out into the market-place, industrial growth in the United States exceeded all previous experience. It changed much more than the existing systems of supply and manufacture. The mass production of machines, the new consumerism, the proliferation of steel-framed buildings, the spread of electric power and telegraph networks – all these had created an appetite for metals that traditional mining methods could not satisfy. The techniques of underground mining had been exploited to their limits. The richer seams of ore had given out, and the costs of recovering and treating the poorer grades were continually rising. There were still vast resources of minerals in the earth, but to recover and use them a new technology was needed. The answer came, as so often in America's experience, from the West.

Barely a year after the Mormon settlement of Salt Lake City in 1847, the California goldrush had brought hordes of fortune hunters pouring through Brigham Young's austere and self-denying domain on their way to the Pacific coast. Despite Young's ban on prospecting for gold (which, he said, was for 'paving the streets of Hell'), many prospectors were inevitably side-tracked into exploration of the mountains which rose steeply to the east and south of the Great Salt Lake. Sure enough,

they found rich veins of gold, silver and copper in a place called Bingham Canyon. As more and bigger discoveries were made in what became the state of Utah, local fortunes began to pile up. But what followed had world-wide repercussions.

In 1887 an observant miner found that an entire mountain in the upper reaches of Bingham Canyon appeared to be impregnated with copper ore, in finely-divided, evenly-distributed particles. At some earlier time a tunnel about thirty metres long had been driven into the mountain, but because the rock assayed only about two per cent copper, which at that time was not worth exploiting, the idea of mining it had clearly been abandoned.

For more than a decade the massive deposit, containing copper worth tens of millions in the sunrise of the electrical age, tempted and frustrated the miners. Eventually a young consulting engineer named Daniel C. Jackling examined the mountain, and put in a report which was to change the landscape of mining all over the world.

What Jackling proposed was a new dimension of the ancient system of open-cut working. It meant creating tools capable of moving mountains. Jackling envisaged huge mechanical shovels, larger than any that had ever been built, working in a single gigantic open pit, loading endless trains of rail-road cars with unheard-of daily tonnages of ore for the crushing mills.

Jackling's system was put into operation at Bingham Canyon in 1903. The pit which first began to eat out the crown of the mountain is now a vast amphitheatre nearly one kilometre deep – the largest man-made excavation on the face of the Earth. It has yielded more than eleven million tonnes of copper metal, worth in excess of six billion dollars, or more than the combined values of the California and Klondike goldrushes and the legendary silver Comstock Lode. Today the trains still move half a million tonnes of ore and rock *a day* out of the deepening pit through tunnels in the walls, and the mine is still profitable, even though the grade of ore is down to less than half of one per cent copper.[56]

But Jackling's concept solved only half the problem of using low-grade mineral deposits. There remained the difficulty of economically extracting the metal from the ores. It was a constraint which was suffered in several countries where large resources of low-grade minerals were waiting to be exploited. Their recovery had never been attempted, because the systems then in use for separating metallic ores from rock were not effective for all metals. In general, such systems relied upon differences in density between metals and rock. They worked by washing the finely crushed ore over shaking tables, whereupon the metals sank and the lighter rock was washed away with the water. But some metals, such as zinc, had the same specific gravity as the rock, and refused to sink. Where the waste went, so went the zinc.

Finally, someone suggested that if zinc could not be made to sink, perhaps it could be made to float, leaving the rock in the water. The unexpected tendency of some minerals to float had been noticed as long ago as 1789, but had never been exploited. At that time, a Welsh bishop, Richard Watson, had observed that if he put powdered lead ore

Opposite: The 19th century developments in electric lighting and power created a demand for copper which conventional mining methods were unable to meet. The giant-scale open cut mining techniques, first introduced at Bingham Canyon in Utah in 1903, made possible the recovery of large deposits of low-grade copper ore. *The dimensions of the Bingham Canyon pit, which is today the largest man-made excavation on earth, can be judged from the apparent diminutiveness of the power shovels working on the terraces; these machines are as high as a four-storey building.*

233

into a glass of water and then added nitric acid, the rising bubbles produced by the reaction carried particles of lead to the surface with them.

A century later, quite independently, a Melbourne brewer made the same observation, only in his case it was that bubbles rising in fermenting beer carried impurities to the surface. In this he was to glimpse the yeast of a new technology in mining. It was a likely accident to happen in Australia, where beer-drinking and mining had become inseparable cultures.

The brewer, Charles Potter, was also an inventor, and he decided to apply his observation to a problem then widely discussed in Melbourne – the huge losses by mines at Broken Hill through their inability to recover all the available silver, lead and zinc from the ore they were extracting. They were getting less than half the silver, about two-thirds of the lead, and virtually none of the zinc. It was only the extraordinary richness of the great Broken Hill lode that made mining there possible.

After a decade's work, Potter developed a process for recovering metals by flotation, very similar to Bishop Watson's long-lost experiment, and he patented it in 1901. It was tried out by one company in Broken Hill with great success. They achieved a recovery figure of about sixty per cent for zinc, as well as a high proportion of the silver and lead. Once the idea of flotation got about, variations were rapidly tried out, for vast sums were at stake. BHP, the mining giant, had tailing dumps containing metals worth more than twenty million dollars waiting for just such a method of treatment. Before long, there was a system which could float different metals selectively, and thus separate them.

Essentially, the flotation process, which was largely pioneered in Australia but soon spread worldwide, involves blowing air into a bath containing finely ground ore and liquids, to produce a rising stream of small bubbles. By adding particular chemicals to the bath, the particles of one mineral become coated with the chemical instead of water. When they encounter an air-filled bubble they alone attach themselves to the bubbles and are carried to the surface, leaving everything else behind. The froth, in this case, is richer than the substance, and easily gives up its metal content. In the words of historian Geoffrey Blainey, 'thus the bubble was cultivated and tamed, taught how to capture some minerals and reject others'.[57]

The flotation process arrived just in time to complement Jackling's innovation in open-cut mining, and at Bingham Canyon there were soon rows of tanks frothing with green bubbles; it was this giant mineral brewery which finally made the low-grade ore worth recovering. And so it was that the industrial nations of the world were able to surge into the twentieth century on a rising tide of production, their raw materials assured by swelling eruptions out of the new craters which began to pock-mark the face of the planet.

But while the Americans were mobilising their vast resources of minerals and energy in that last quarter of the nineteenth century to launch the world into the Machine Age, the Europeans were laying the foundations for an even more radical transformation. *Their* resources were the intellectual powers of their scientists.

In the quiet university towns of Cambridge and Gottingen, in labora-

234

tories in Paris and Copenhagen and Rome, and in the recesses of the human imagination, a wave of creative energy was forming. Over the next few decades it would carry science to the threshold of comprehending the most profound mystery in the physical universe – the nature of the atom, and of the awesome forces locked up within it.

Like many significant advances in science, the revelation began with the observation by the receptive human mind of some anomalous behaviour in nature. In this case, it was the fact that some metals behaved differently from others – they emitted a form of energy that no one could put a name to. The first steps towards identifying and understanding the nature of that energy were taken in Paris.

Today the Curie Institute forms part of a complex of grey stone buildings in the heart of Paris, in the clutter of narrow streets near the Sorbonne. The institute itself occupies the ground floor of a plain brick building, and consists of two small laboratories, two offices and a hallway serving as a museum. The corner office looks out on to a quiet courtyard, deeply shaded with trees in summer, with one or two garden seats arranged at either end. In one corner of the courtyard there is a stone plinth bearing the bronze busts, dark with lichen and patina, of a stern couple – Pierre Curie, a French physicist, and his Polish wife, Marie.

The name Curie inevitably brings to mind the term radioactivity. It is one of the most contentious words in our language. To many it has frightening implications of nuclear fall-out, reactor melt-down, the long-term contamination of the atmosphere, the threat of radiation sickness, death from cancer, and genetic damage to generations yet unborn. To others it means powerful weapons against cancer and other diseases, renewable energy resources to replace those dwindling oil reserves and stave off the menace of acid rain and the 'greenhouse effect', and a whole new range of scientific and industrial tools.

What is beyond dispute is that the discovery and identification of radioactivity in metals such as radium and uranium changed the world irrevocably. The term radioactivity was coined by Marie Curie, who worked in the Curie Institute from the end of World War I until her death in 1934. She spent many reflective hours, in the last years of her life, in the shady retreat outside her office. One of the last things she did was to plant the white climbing rose which still flowers each spring at the edge of the courtyard.

For sixty years two generations of Curies, by their exploration of the almost unimaginable potential of the radioactive elements, played a central role in the development of modern nuclear physics. Pierre and Marie, their daughter Irene, and their son-in-law Frederic Joliot all received Nobel prizes. Of course, scores of other scientists from many nations were involved in the long, sustained thrust of the intellect which in the end forced open the mighty doors to the heart of matter. But that is a story much too wide-ranging to be dealt with adequately here, except to acknowledge the seminal role played by the Curies. It was their insight and their work with those mysterious and unsuspectedly powerful radioactive metals, as much as anyone else's that laid the foundations of the Atomic Age.

Marie Sklodowska left Poland in 1891 to enter the Sorbonne in Paris.

Marie Curie, who coined the term radioactivity, died in 1934, a victim of the great forces that she helped to liberate. *Marie Curie's laboratory at the Curie Institute in Paris, where she spent the last years of her life, has been preserved just as she left it.*

She obtained her Bachelor of Science degree in 1893, and was working as a laboratory assistant when she married Pierre Curie in 1895. Soon after, while deciding upon a subject for her doctoral thesis, Marie became intrigued by Henri Becquerel's demonstration of unseen but apparently powerful 'rays' emanating from salts of the metal uranium.

Becquerel had been experimenting with X-rays, discovered in 1895 by William Rontgen. X-rays were produced by certain crystals when exposed to ultra-violet radiation, and had the power to penetrate substances opaque to ordinary light. In trying various crystals, Becquerel observed that uranium salts produced penetrating rays which cast a 'shadow' of a metal object on a shielded photographic plate – even without being excited by ultra-violet energy. Becquerel described this emanation of uranium as a kind of unique metallic 'phosphorescence'. Marie Curie began to search for other elements with similar properties.

Marie eventually identified the same kind of emanation from the metallic element thorium, and gave it the name 'radioactivity'. Pierre joined Marie in her work, and discovered that radioactivity consisted of flying packets of energy, or particles, with different kinds of electrical charges – positive, negative and neutral. The great New Zealand physicist Ernest Rutherford was later to name them alpha, beta and gamma rays. At the Curie laboratory in Paris the nucleus had begun to give up its secrets.

Next, using apparatus of undeniable crudity (now preserved in the Curie Institute laboratory), the Curies were able for the first time to measure the intensity of this radiation. In 1898 they announced the discovery of two new radioactive metals, radium and polonium – the latter named for Marie's Polish homeland. For their work both Pierre and Marie, with Henri Becquerel, were awarded the Nobel prize for physics in 1903. And so, by the beginning of this century, the doors into the inner world of the atomic nucleus had been pushed open, just a chink.

In 1906 the partnership was tragically broken, when Pierre was killed in a street accident in Paris. Marie carried on the work, and took up Pierre's professorship at the Sorbonne, becoming the first woman to hold such a post. In 1911 she won a second Nobel prize, this time for chemistry, for the discovery of radium and polonium, and for the isolation of pure metallic radium.

Then came World War 1, and Marie and her daughter Irene worked in the French army medical service. They explored the use of X-rays for the assessment of bone damage and for other diagnostic purposes. Throughout those years, unsuspecting, they both undoubtedly received large doses of X-rays. The hazards of such radiation were little understood at the time, and the kind of shielding used today was virtually unknown.

At the end of the war the French government created the Radium Institute in Paris to build on the Curies' work, and Marie joined the staff. She brought her equipment from her old laboratory, moved into the corner office of the building which was eventually named after her, and resumed her work. By this time Irene had become involved in atomic research, and joined her mother at the Institute. In 1925 the two Curies were joined by a young assistant, Frederic Joliot. Frederic was an engineer, and Irene guided him in his rapid mastery of laboratory research techniques with radioactive materials. In 1932 they were married, and in partnership worked towards that slow assembly of our picture of the atom.

Frederic and Irene were particularly interested in the structure of the atomic nucleus being revealed by the work of Rutherford at Cambridge. At that time there were no 'atom-smashers' for probing the make-up of matter, but Frederic and Irene had the immense advantage of a large stockpile of highly-radioactive material that had been accumulated for research by Marie Curie. Using an intensely radioactive polonium source, they bombarded a thin sheet of aluminium with alpha particles. Neutrons and other particles were knocked out of the aluminium nuclei, leaving short-lived traces of a different element, radioactive phosphorus. Frederic and Irene had realised the dream of the alchemists – they had produced the transmutation of one element into another.

That same year Marie Curie died, her body blanched by her lifelong involvement with the great forces that she helped to liberate.

In 1935, Frederic and Irene were awarded their Nobel prizes for the synthesis of new radioactive elements, and went on with their experiments in bombarding nuclei with neutrons. In 1939, Frederic Joliot was able to demonstrate the process of nuclear fission, which had been predicted by the three German physicists Otto Hahn, Lise Meitner and Fritz Strassman. This development work was critical in the calculations which were to produce the first chain reaction in 1942, the nuclear power reactor, and the atomic bomb.

Both Frederic and Irene were excluded from the Allied nuclear war effort because of their affiliations with the French Communist Party, but they were the creative force behind the establishment in 1945 of the French Atomic Energy Commission. This enabled France to break the Anglo-American monopoly of nuclear technology after the war, and to adopt her controversial independent nuclear policy.

Like her mother, Irene finally became a victim of her great appetite for research, and died of leukaemia in 1956. Frederic died two years later. In the courtyard of the Curie Institute, however, when the white rose that Marie Curie planted blooms beneath the trees where she and her daughter walked, those momentous years around the turn of the century – the dawn of the Atomic Age – do not seem far away.

That period of fundamental discovery was the climax of a century in which the scientists had given mankind, in geology, a new tool for unearthing the resources of the planet; in aluminium, a miraculously light metal for the conquest of the air; in electricity, a uniquely versatile source of energy; and in the keys to the atomic nucleus a potential which would become a central and dominant dilemma of the twentieth century.

CHAPTER 11
The age of metals; can it last?

Like a stranded space station, gleaming with lights and humming with energy, the Pompidou Centre rises improbably above the grey slate rooftops in the heart of Paris. Free from the conventional skin of masonry or brick, each facade of the huge structure is a startling collage of brightly painted metal tubes, struts, ribs, plates and port-holes. Escalating slowly across its front elevation in diagonal glass tubes, its ceaseless intake of visitors shows up like a barium meal. Here is a visceral display of function in architecture – the anatomy of the twentieth century exposed. It speaks for our time as the Crystal Palace spoke for the nineteenth century.

Although the Pompidou Centre is dedicated to the display of French culture, and contains conventional auditoriums, offices, restaurants and galleries, in all other respects it is a point of departure from the traditions which surround it. It was not built by craftsmen, laying stone upon stone, but assembled by machines, from steel and aluminium components made elsewhere by other machines. In its confident rejection of traditional building materials it argues that metals, which have at some time displaced natural staples in almost every artefact that man uses, are now ready to provide his habitations.

For some, the prospect of an all-metal world is not only logical but exciting. For others, it is inhuman. But whether the Pompidou Centre turns out to be a forecast or a folly will depend upon answers to questions of the present: Are there sufficient metals on our planet for all who want them? Will society accept the environmental penalties of producing metals on an ever-increasing scale? And what will we use for energy if *all* the world is to be carried foward into the Age of Metals?

For ten thousand years, metallurgy has accompanied the march of man, transforming his way of life and filling his material universe. The place of metals in modern civilisation is now beyond question or choice, despite the increasing versatility of plastics and ceramics. And yet, for all its high technology, metallurgy has never quite lost touch with its ancient origins.

The Pompidou Centre in Paris is a visceral display of function in architecture. In its metal structures and components many see a glimpse of a future in which metals, having at some time replaced natural materials in almost every artefact that man uses, will also provide his habitations. *The Pompidou Centre is devoted to the display of art and culture.*

The passenger jetliner is still barely thirty years old, but it is among the highest reaches of the art of man the metalsmith. Among the dozen or more metals used in its construction, the four tonnes of exotic elements in each engine are critical – nickel, titanium, chromium, cobalt, niobium and tantalum. Yet the very heart of this new technology – the turbines in its fiery core – are made by one of the most ancient ways of working metal, lost-wax casting.

The origins of lost-wax casting are unknown, but it eventually developed independently in many parts of the world. It was used in the Near East as early as the fourth millennium BC to make small figures or trinkets. That original technique is still taught as part of the sculpture course at the Tokyo University of the Arts. As well as creating works of their own, the students cast replicas of rare and ancient works of art in bronze, which were themselves made by this technique. A typical example is a bronze hand mirror of the Heian period, dating from about AD 900, and now in the Emperor's collection in the Imperial palace. The reflecting face is smoothly polished bronze but the back carries a raised, decorative pattern of flowers and birds.

The process begins with a wax model of the object to be cast. In ancient times, this was shaped by the craftsman from beeswax. Today, to copy something which already exists, like the mirror, a coat of liquid latex is applied. When this is peeled off it forms a mould, which is filled with wax to produce the model.

Soft clay, applied in layers, is used to form a mould around the wax model of the mirror – first a pasty liquid to fill all the contours of the model, and then increasingly coarser material to provide strength to the mould. A small hole is left in one end of the shroud, extending through the clay to the model. When the mould is heated in a furnace, with the

LOST-WAX CASTING

One of the oldest ways of working with metal is lost-wax casting. At Tokyo University of the Arts, students learning to work with bronze make replicas of objects created originally by this ancient process, using exactly the same materials and techniques. Here a bronze mirror of the Heian period, around AD 900, is being copied. *Below*: The process begins with a wax model of the original, which is in the Emperor's collection at the Imperial Palace in Tokyo. *Right*: The model is coated with liquid clay – the first stage of creating a mould around it.

Left: Many coats of clay are applied, using increasingly coarser material to build up the strength of the mould. *Above*: When one side of the model has been coated to the desired thickness, it is turned over and the other side treated similarly. In the process, a small channel is left connecting the handle of the wax model with the outside. Through this the wax will eventually be 'lost'.

The finished clay shroud is further strengthened with a loop of iron rod before baking. The coated clay model is placed in the kiln, made of mud bricks, with the opening leading to the handle pointing downwards. During the baking the wax runs out through the hole, and the clay is fired into a rigid shell.

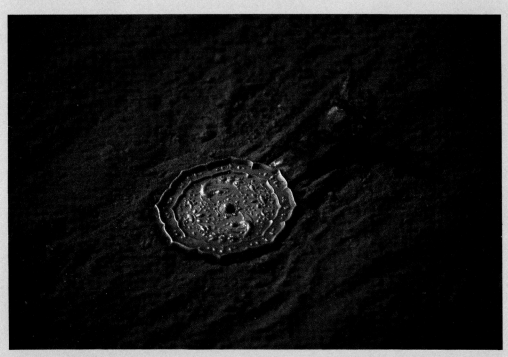

Above: The baked mould now contains a void which exactly corresponds to the shape of the 'lost' wax model of the mirror. It is filled through the hole with molten bronze and left to cool. *Left*: The finished bronze casting, when broken out of the mould and cleaned, is an exact replica of the original, in execution as well as in appearance.

hole pointing downwards, the wax melts and runs out – is 'lost' – while the clay shroud is baked hard. It now contains a void which exactly corresponds to the shape of the original wax model.

The bronze is prepared by melting copper and tin together in a crucible, in the classic proportions of ninety per cent copper to ten per cent tin, and is poured into the mould through the hole. With complicated pieces, several pouring channels or 'sprues' may be included in the mould. When the bronze has set, the clay mould is broken open, the stem of metal which has solidified in the sprue is cut off, and the join smoothed down. When cleaned with a wire brush, the casting emerges as a perfect replica of the original Heian mirror in form, content and execution.

Half-way across the world, in the United States, the Howmet Turbine Company uses exactly the same process to manufacture the turbine blades which drive the engines of two-thirds of the world's jet aircraft. For all the advances in metallurgy in the past thousands of years, only lost-wax casting can give these vital components the strength to survive the ferocity of their working environment.

Few artefacts in constant everyday use are subjected to the stresses imposed on the turbine blades of a jet engine. From a cold start they rapidly reach operating temperatures of well over 1000°C, followed by numerous temperature changes as power is varied during flight. The blades rotate at eight to ten thousand revolutions per minute, with centrifugal forces trying to tear them apart. Yet they must not bend or deform from their shape in the slightest, while operating non-stop for perhaps fourteen hours in a blast of corrosive gases hot enough to melt any ordinary metal.

The casting of turbine blades, as with the Heian mirror, begins with wax models. These are produced by injecting synthetic wax into moulds under pressure. The models are finished by hand to remove the injection stems and obtain a perfectly smooth surface on the wax. The wax blades are then assembled in clusters on a base plate, picked up by a robot arm, and dipped like chocolates in a bath of thick brown liquid. This is a slurry of fine ceramic material, similar in function to the first coating of liquid clay applied to the Heian mirror model.

From the liquid bath, the robot swings the dripping blade models into an alcove, where they are showered with ceramic sand-like particles. This encrusts the liquid coating, forming the basis of a mould. The dipping and showering sequence is repeated several times, until a thick, firm ceramic coating has been built up around the wax blades. Like the mirror mould, each of the blade shrouds has a small opening in its base. When they are heated in a steam autoclave, the wax melts and runs out. The moulds are then fired in a gas furnace into rigid ceramic shells. These are filled with molten metal to produce finished turbine blades, to extraordinarily exact dimensions.

The only real difference between the casting of a Heian mirror and a turbine blade is the metal used. Both are classed as alloys, but that is where the resemblance ends. The so-called 'super-alloys' used at the Howmet plant are sophisticated combinations of metals with particular characteristics, tailored to meet the almost impossible stresses of the jet turbine. Their major ingredient is nickel, for strength and hardness, but a typical super-alloy includes chromium, cobalt, molybdenum, tungsten, tantalum, aluminium, titanium, boron and zirconium.

Casting with super-alloys is an exacting scientific process. Some metals, such as boron or zirconium, may be included in small but remarkably precise proportions – one part in twenty thousand, or 0.005 per cent. Others, such as aluminium and titanium, react with air, and castings containing these metals must be made under high vacuum. Most importantly, the manner in which the super-alloys 'set' is critical to their ultimate strength.

When molten metals solidify, they develop a lattice-work of crystals, and the size and alignment of the crystals dictate many of the physical properties of the metal. Crystal formation in a super-alloy casting can be influenced by many factors, even the composition of the surface of the mould in contact with the molten metal. After several decades of manipulating crystal alignment in their castings, the Howmet scientists have learned how to make the molten metal in the mould set into a *single* crystal, the size of the whole blade (which can be up to twenty centimetres long). This is close to the ultimate level of technology, in one of the most arcane reaches of metallurgy – but it is still dependent upon the ancient art of lost-wax casting.

Replicas of objects like the Heian bronze mirror from the Imperial palace in Tokyo go to museums. Like their ancestors, they may last for a thousand years. The turbine blades in a jet engine will survive for only a few thousand hours. But in that time, all over the world, millions of lives depend upon them.

In the long journey from simple shapes cast in clay to intercontinental supersonic travel there have been many spurs and impulses, but one unflagging stimulus has been the hunger for liberation by wealth. It has driven civilisation to the very edge of the known world, in a never-ending and sometimes heedless quest for the shining prizes hidden in the Earth.

One of the world's far corners, and one of its wilder shores, is the west coast of Tasmania, where great swells break after their long run before the 'Roaring Forties' across the Southern Ocean. The misty, desolate beaches and the black, wave-drenched cliffs are as unmarked now as when Abel Tasman first glimpsed and cautiously skirted them in 1642. There are no towns or permanent settlements on the west coast, and the hand of man is nowhere to be seen except in the lonely, isolated huts of fishermen.

The hills behind the coast have not escaped so lightly. In the closing decade of the nineteenth century, rich deposits of copper were found beneath the dense wet forests which clothed the south-west of Tasmania. The outcome was the mining community at Queenstown, and above it the present lunar landscape of barren red and yellow ridges, dotted with blackened tree stumps and eroded outcrops of rock.

First, the hillsides around Queenstown were cleared of timber to feed the furnaces down in the valley. The fumes from the smelters killed off the rest of the vegetation, and prevented natural regeneration on the slopes. The heavy rainfall did the rest, scouring away the topsoil. The result was a supremely awful man-made scar, which represents the more unthinking excesses produced by man's voracious appetite for metals.

In the harshness of this landscape, many would see that ominous conjunction of events which invokes the popular nightmare of a planet made, in the end, uninhabitable by the demands of the consumer society, and its stimulation of what are considered to be unnecessary needs and wasteful appetites. Few, however, yet choose to divest themselves of their metal-dependent way of life. It is an ambivalence which has accompanied mining and metal-winning throughout the thousands of years that mankind has been using metals. It is inevitably sharpened by the increase in scale, and the intrusion of what was once a remote and largely unmonitored activity into the consciousness of the city dweller. It is part of the price that mankind has chosen to pay for its passage from the Stone Age into the Age of Metals.

It is ironic, too, that the sheer extent of devastation near Queenstown has made it, for some, a tourist attraction, its sulphurous genesis somehow sanitised by time. There have even been serious suggestions among the population that any trees found regenerating on the slopes should be removed – to preserve the stark and colourful backdrop to a town where mining is no longer the sole reason for existence. In any case, it is probably true to say that what happened around Queenstown at the beginning of the century could not happen now. Current attitudes make it unacceptable, and modern technology makes it unnecessary.

However, the industrialisation of human society, and the stimulation of its appetite for minerals, has sharpened the conflict between needs and costs. That conflict is a volatile political reality in the densely populated urban sprawl of Western Europe – but there, environmental concerns have produced impressive community solutions.

West Germany obtains one-third of its electricity from a chain of huge coal-burning power stations. The largest of them rises above the green fields near Cologne, its tall smoke-stacks and squat cooling towers dominating the flat landscape. It consumes more than one hundred million tonnes of brown coal a year, or nearly two million tonnes a week. This fuel is mined by open-cut from a massive coal seam beneath the flood-plain of the Rhine.

To get at this fuel, buried under two hundred metres of overburden, the Germans have developed the largest mobile machines ever built. Weighing up to thirteen thousand tonnes, these gargantuan mining tools move on crawler tracks. Overall, they measure more than two hundred metres in length and are seventy metres high. An enormous revolving bucket-wheel at the end of a long boom scoops up overburden or soft brown coal and dumps it on to conveyor belts running the length of the boom. Other belts carry the material away – the overburden to be stored, the brown coal to be made into briquettes for the power stations. Altogether, more than a million tonnes of material a day is removed from the open-cast pit. It is the largest earth-moving operation ever mounted.

What makes the Rhine brown-coal project remarkable, however, is

not just its dimensions, but the fact that this vast gouging and overturning of the landscape is taking place in the heart of what is still the richest farmland in Germany. The flood plain of the Rhine is a deep bed of loess – fine soil accumulated from wind-blown dust, and enhanced organically by careful mixed farming and animal husbandry since Roman times.

However, the coal which lies beneath the patchwork of farms and villages is inseparable from Germany's industrial pre-eminence in Europe, and will continue to be so until well into the next century. A conflict of interests between town and country is therefore fundamental and unavoidable. But here it is being resolved in a harmony of opposites – total evisceration of the landscape is followed by its complete restoration, of villages, churches, farms, roads, lakes and forests. It is not just the landscape itself which is preserved in this way, but the whole fabric of the rural society which has existed here for centuries. This has been achieved by the exemplary co-operation of the people, the mining company and the state government.

Plans are agreed for each area long before the huge machines move in. The inhabitants of the doomed villages and farms move into new villages and new farms, built for them on land which has already been restored from previous mining. The people buy their new properties

In Germany the largest land machines ever built are used to gouge brown coal out of the rich farmland of the Rhine valley for use in power stations. *The environmental scars are completely erased by an extraordinary programme of total restoration of land, forests, lakes and villages, which is a model for mining projects everywhere.*

with the compensation paid for the old, which simply disappear as the new open-cut expands. When the coal has been extracted – a matter of a few years – the vast hole is filled in and the landscape restored, ready to be re-occupied by people removed from somewhere else.

Where farmland is involved, the mining company stores the valuable loess until eventually the open-cut has been filled in and levelled. The soil is then mixed with water and pumped out into shallow 'polders' or ponds. When the water evaporates it leaves a thick deposit of silt. This is repeated until the required depth of loess is obtained. If necessary, the mining agronomists add organic material to reconstitute the soil's fertility before selling it back to the farmers. The process is so successful that reconstituted land now sells for more than the original farmland. The farmers particularly appreciate the fact that their new farm buildings and farmhouse are built right on their new land, to their own specification. This guarantees the most efficient daily use of human and mechanical resources.

More than twenty-five thousand people and scores of villages have been moved and resettled in the Rhine brown coal zone. The entire cost is borne by the mining company, with the state government providing funds for improved civic services, such as schools. In the end, the cost is built into the price of brown coal to the power stations, but for West Germany it is still cheaper to obtain its energy this way than by buying oil from OPEC or from Britain's North Sea oilfields.

There are also general environmental gains for the whole Cologne-Dusseldorf region, one of the most heavily built-up regions in the world. The inevitable blight of two centuries of heavy industry has been softened, and living made more pleasant for the inhabitants of this grey, smog-blanketed valley. One fifth of the restored land is landscaped and planted with trees to create woodland, streams and winding country lanes, where people now walk and cycle. These areas are now among the most natural looking environments to be found in Germany, where very little original forested land remains.

At the southern end of the coal-bearing zone, where open-cut mining began in the 1930s, and where the policy of total restoration was first introduced, the deception is complete. Here mature forests come right down to deep lakes, where people sail and canoe and swim in pristine surroundings. Among those who seek tranquillity here there are few who can still remember when the lakes and forests were one vast hole in the ground.

The significance of the Rhine brown coal operation is the example that it offers and the message that it conveys: there is no environmental damage caused by man's need for minerals that cannot be repaired, if society is prepared to share the cost. That reality was not always faced in the past, and has still not been confronted in many parts of the world. It took the coincidence of economics, politics and survival to obtain acceptance in West Germany, and later more widely.

There is of course another way of containing the impact of mining. It is to make less profligate and more lasting use of the metals already won and processed. Scrap steel, aluminium and copper, in particular, are being recycled in greater quantities than ever before. One obvious saving is in energy. The reprocessing of scrap aluminium and copper takes only a few per cent of the energy required to smelt the original metals, while scrap steel needs only fifteen per cent of the original energy to be re-melted. Where precious metals are concerned there is an even greater incentive to recycle them, especially as they are resistant to oxidation and corrosion and therefore lend themselves to recovery. The outstanding example is silver, and the quite remarkable campaign of recycling which is carried out by the world's largest user of this metal – Kodak.

Silver was one of the first metals to come to the attention of early man, and it was greatly admired for its lustrous character. In its native form, silver was quite uncommon, and in some early societies, such as the Egyptian Old Kingdom in the fourth millennium BC, it was more precious than gold. As the metalsmiths learned how to smelt it from the reasonably abundant silver-lead ores, silver came into much more general use, especially for coinage. However, its most unusual and specialised role only developed during that great burst of applied technology in the United States towards the end of the last century.

One of the most characteristic examples of the American instinct for the popularisation of invention was devised in 1880 by a bank clerk named George Eastman. At that time, photography of people required the subject to go to a studio and sit with his or her head locked into a kind of Iron Maiden device for perhaps twenty minutes, for a single portrait faintly etched on a metal plate. Making photographs outside the studio meant carrying around a cumbersome wooden box camera, and coating heavy glass slides with wet chemicals before exposing them. The plates then had to be developed on the spot with yet more volatile and messy liquids.

Photography – the industry which has become the world's largest consumer of silver – was pioneered in Europe but given its greatest impetus to growth by the inventions of an American bank clerk, George Eastman. *This early self-portrait has Eastman's own comments on the print quality written across it.*

Because of such technical requirements, photography was effectively restricted to professionals, some of whom, in Europe in particular, had raised the new means of recording reality to an art form. George Eastman believed that there ought to be an easier way of making photographs. He also fervently believed that this new technology should be made available to all, for use as a kind of scrapbook of family life.

Eastman had been provoked into this opinion by his first experience with photography. He had planned to take some pictures on his annual holiday, but when confronted with the pile of apparatus, plates, chemicals and utensils that were involved, he cancelled the vacation and spent the time in his mother's kitchen, seeking a simpler system. In the end it took him several years, for he had to teach himself not only chemistry but a great deal about mechanics, physics, optics and tool making.

What Eastman finally put on sale was simplicity itself – a small, light box with a simple lens set in one end, and a rudimentary shutter operated by a button on the top. What went inside the box, and replaced the glass slides still in universal use, was revolutionary. It was a roll of the newly-discovered plastic, celluloid, already coated with chemicals. Eastman had worked his way to this concept of roll film through dry-coated glass slides and then coated sheets of film.

Eastman's camera was sold with a roll of film inside. It was long

Left: Eastman's first photograph with his sheet film, which he developed from the cumbersome glass slides then in general use. *Above*: Eastman, on a cruise with the first Kodak, foreshadows the worldwide mania for taking snapshots which his invention will soon arouse.

enough for one hundred exposures or 'snapshots', as they became known, on the analogy of a swiftly fired gun shot. When the roll was finished, the user sent the whole camera back to the factory. There the film was developed, the prints mounted, and prints and camera, with a fresh roll of film in place, were returned to the owner – all for the price of ten dollars.

Before Eastman's invention, photography was beyond both the capacity and the patience of ordinary people. Because of him, it became a universal folk art. His advertising slogan, 'You push the button and we do the rest', was among the more compelling invitations in the history of merchandising. The name Kodak was one he made up. He wanted, he said, a word which 'could be pronounced in any known language'.

Thus George Eastman launched a craze which became the most popular and widely practised hobby on Earth. And because it all depended upon silver nitrate's unusual property of turning black when exposed to light, photography eventually became what it has remained – the world's largest consumer of silver. The Kodak company uses half the total amount of silver circulating annually in the United States, and more than many countries, including Germany and the United Kingdom.

Side by side with its enormous production of photographic products,

Far left: Eastman soon followed up the Kodak camera with the Box Brownie, which brought popular photography within reach of almost everyone. *Left*: Eastman's company rapidly grew into what it has remained – the world's largest consumer of silver. Kodak today uses half the silver circulating in the United States, and more than Britain or Germany. *Historic apparatus and photographs can be seen at Eastman House, where Eastman lived in Rochester, NY.*

Kodak conducts a never-ending salvage operation to retrieve the precious metal. Exposed materials are collected all over the United States and shipped to its main plant at Rochester, New York. It is a simple matter to reduce waste film to ash in a furnace, whereupon the silver runs free in a pure molten stream, and can be cast into ingots ready for use again. From fleeting images frozen in snapshots, old movies and X-ray films, Kodak recovers more than six hundred tonnes of silver a year – worth hundreds of millions of dollars.

An increasing proportion of the silver used today goes into the electronics industries, because it is an incomparable conductor of electricity and heat. Thus the world demand for silver continues to climb. The steepness of the upward curve can be judged from the fact that of the million tonnes of silver mined in the past five thousand years, more than half has been produced in the past hundred years.

The production of many other metals has also risen dramatically to meet world consumption, which this century has been doubling about every twenty years or less. Demand has been stimulated by the realisation that, in contrast to other commodities, the cost of most base metals has been falling, in real terms, through more efficient methods of mining and processing. However, despite the still-growing appetites of industry in the Western countries, and the spread of industrialisation to the developing nations, geologists are convinced that the fear of running out of metal resources is illusory. In fact, that prospect has never been less likely than it is today.

The historic argument is persuasive – demand has always fostered innovation in search and discovery. The current advances in this field are providing mineral prospecting with facilities which would have seemed inconceivable even twenty years ago.

The expansion of the horizons of the earth-trudging prospector to the perspectives of the orbiting geologist in less than a hundred years is an extraordinary accomplishment in itself. Standing now outside the Earth itself, surveying it from the all-encompassing viewpoint of the satellite, the mineral seeker has a powerful battery of modern divining rods to help him see beneath the old and wrinkled skin of the planet – a whole spectrum of electromagnetic pulses, from infra-red to ultra-violet. More recently, the known capacity of electrical and magnetic probes to find hidden bulls-eyes of metals has been augmented by the enormous potential of satellite imaging, through the use of the Landsat satellites in orbit around the Earth.

Photographs from space, made in various combinations of light waves, with their detail and clarity enhanced by computer technology, can now clearly differentiate between geological features which would be indistinguishable in ordinary photographs. The light reflected by plants in many cases can also indicate the chemistry of the soil in which they are growing. Such clues may be sufficient to identify the underlying rocks, although totally hidden from view. This evolving technique is already being used in the selection of areas for ground exploration. One important deposit that it has helped to prove is a vast bauxite province in the remote and rugged plateau country of the Kimberleys, in north-western Australia. To our still spotty and incom-

Left: New techniques of finding minerals are more than keeping up with demand, so that the danger of the world running out of metals is less likely than it has ever been. *This is an example of computer-enhanced 'satellite imaging', which has helped to establish the extent of a vast new deposit of bauxite in the Kimberley region of north-west Australia. In these 'false colour' images the usual colours are reversed. The bright red patches are not bauxite but green vegetation. However, this is a guide to the presence of bauxite, because plants grow well on the deeply weathered bauxite and laterite soils which have developed on the volcanic rocks in this region. For example, Cape Bougainville (top) is covered with bauxite, and shows up as well vegetated. Sandstone, which bears no bauxite and forms poor soils, supports little vegetation, and therefore shows up in pale blue tones. Below*: The enormous iron ore deposits of the Pilbara in Western Australia, being mined at the rate of hundreds of millions of tonnes a year, are one of the reasons why the world's centre of power is shifting from the Atlantic to the Pacific basin. *A large blasting sequence at Mt Tom Price.*

plete picture of the vast Australian land mass, Landsat imaging is adding 'scenes of grandeur, delicate beauty and complex patterns that pass beneath the satellites each day . . . Dreamtime art, painted on a huge natural canvas by wind, sun, water and time'.

What emerges from each new discovery, however, is confirmation that although the total of mineral resources on the planet must obviously be finite, we have still little idea of their true magnitude or indeed where those resources lie. Australia, the last of the inhabited continents to be explored for minerals, has even now hardly been scratched. The enormous lode of uranium, copper and gold located deep beneath a sandstone capping at Roxby Downs had yielded no hint of its presence to conventional exploration techniques. Many more such riches may lie beneath the vast Australian emptiness.

It is only since World War II, for example, that one of the oldest landscapes in existence, the Pilbara in Western Australia, has been closely examined, and found to contain vast beds of iron ore laid down on the floor of ancient seas. And even since these resources began to be opened up, at mines like Mt Tom Price and Mt Newman, Australia's reserves of iron have been proved to be at least one hundred times greater than those upon which both national and international economic policies were based at that time. Such discoveries of iron ore, together with subsequent finds of copper, bauxite, uranium, nickel, oil, natural gas, oil shale and diamonds have confirmed Australia's transformation from a predominantly rural economy into a resource-based nation. The fleece which kept Europe warm and made Australia rich is being outweighed by the minerals which feed the furnaces of Japan – and keep Australia rich.

Throughout history, local mineral wealth has made first one nation rich, and then another. Athenian Greece defended itself from the Persians and financed its 'golden age' with the silver it mined at Laurion. Alexander the Great funded his conquests with the gold of Macedon. Imperial Rome armed and paid its legions with the silver and copper of Rio Tinto. In the revival of Europe from the Dark Ages, Germany's mines laid the economic foundations for the revival of confidence and learning which led to the Renaissance. Britain's pre-eminence in the Industrial Revolution was based on her coal, iron, copper and tin. And the far greater resources of the United States – more than twice the mineral production from all previous history – made her the world's most powerful nation.

Australia is a comparative latecomer to the ranks of the major resource nations. One result is that its privileged insulation as an island continent is breaking down. Today the global village is less disposed to look upon nature's haphazard distribution of mineral wealth as a lottery, with big prizes for the few winners who have profited by accidents of geography or colonial enterprise. Access to raw materials has always been a historical envy. Today it is vital to the advanced countries, and is becoming so for those, such as the newly emerging nations of Asia, who are entering the first stage of their own Industrial Revolution.

For the Australian people, the occupation of a treasure house of minerals is likely to prove the most formidable responsibility in their history. They also face the special challenge of devising a workable and equitable compromise between the economic imperatives of white society and the inalienable rights of the original inhabitants of the continent, beneath whose traditional lands so much of the mineral wealth lies waiting.

There are other concerns, which are self-evident at any of the huge mining operations across Australia. The miners themselves live on and off the job in air-conditioned comfort undreamed of by all their predecessors in history. But the plant and the machines assume inhuman proportions, dwarfing the individual in their scale and dimension. There are many who believe that this kind of gigantism must in the end extinguish human enterprise and skill, and limit the freedom of the citizen to manage his life. There is growing resistance to the prospect of all nature reduced to mechanical servitude in the name of material progress. This view overlooks, perhaps, that other servitude which once enslaved all mankind, and still grinds down the greater part of it – the tyranny of daily labour in the fields for bare survival. (The adjustment of human existence to the avalanche of technological change which has swamped it in the past two hundred years may well be the central challenge to our society, but it is beyond our scope here.)

In the meantime, as the mineral resources of Australia flow to the furnaces and factories of the developing nations, impatient for the alternatives to human labour which the West has enjoyed for centuries, it is possible to sense a significant change in balance, as profound as any in human history. It seems apparent that, largely consequent upon the exponential growth of Australia's mineral potential in the last few decades, a vital centre of gravity in the world may be shifting. As it once moved from the Near East to the Mediterranean, from there to western Europe and the Atlantic nations, so now it is moving on again, from the Atlantic to the Pacific. Just as those once-great mines of Spain sustained the ancient civilisations around the Mediterranean, so now around the great rim of the Pacific, from Japan to South-east Asia, to Australia, to South America, and to the New West of the United States, we are seeing the rise of a new sphere of vitality, influence and power.

Australia's raw materials and the industrial heritage of the United States are fused together by the binding force of Japan's ferocious and yet disciplined economic ambition in the nucleus of a new historical prospect – the Pacific Century. The driving energy of this conglomerate is the manpower and the brainpower of Asia.

In an unrivalled synthesis of all previous technology, Japan stands upon a new plateau in the mechanisation of work. The application of robots to the manufacture of everything from cars to watches represents as big a jump from Henry Ford's assembly line as that was from Britain's Industrial Revolution. Like the assembly line, the robot factory has been criticised for its disregard for human values – but it also foreshadows the eventual release of human workers from many of the dangerous, arduous or simply tedious and repetitive jobs that still have to be done, in underground mining, metal foundries, chemical works, factories and harsh natural environments.[58]

In Japan the mechanisation of work has reached levels undreamed of elsewhere. No other nation, perhaps, would consider it possible or practicable to make watches by robot on an assembly line, right down to the connection of the silicon chip 'brain' with filaments of molten gold. *Supervisors at a Seiko plant watch, but rarely interfere, as miniaturised robots assemble quartz watches.*

The progressive replacement of human labour by machines has one other and perhaps even more profound consequence. It means that industry is becoming increasingly dependent upon energy to take the place of people. Already the mining, smelting and processing of the metals which make modern civilisation possible account for one-third of the world's total consumption of energy. Nearly all this industrial demand is met from non-renewable resources – coal, oil and natural gas. The threat of energy starvation has become the fifth horseman of the Apocalypse.

No nation is more conscious of this threat than Japan. Since the rise of OPEC and the 'oil shock' of 1973, the Japanese steel industry in particular has sought protection from that sudden vulnerability in an obsessive campaign to conserve energy. The NKK steel plant at Keihin, south of Tokyo, is perhaps the most advanced example of energy control and conservation in the world today. It foreshadows the way others will have to go, if mankind's current appetite for metals is to be sustained.

The Keihin project began in 1969 as a renovation of three existing steel plants, but the decision was taken to create instead an entirely new plant on a man-made island, one kilometre offshore in the bay. The object was to make it a closed system, for both environmental protection and energy conservation. When completed in 1979, the plant had two of the largest blast furnaces ever built (each producing eight thousand tonnes of pig iron a day), and a complete range of oxygen steel furnaces and rolling mills. Total production is three million tonnes of steel a year. In the planning of its layout to achieve maximum efficiency, and in its automation, this plant is far ahead of any other. What makes it of particular interest, however, is that it is self-sufficient in one important form of energy, electricity.

A thermal power station in the plant supplies all the necessary electricity, and is fuelled entirely on by-products from the operations of iron and steelmaking. The waste gases produced by the blast furnaces, coke ovens, oxygen furnaces and other facilities are not discharged into the atmosphere, but are collected and piped to the power station to be burned as fuel to heat the turbine boilers. The actual pressure of gases in the top of the blast furnaces is also used to drive generating turbines directly to produce power. Waste heat from the rolling mills generates steam to produce even more power.

As far as environmental controls are concerned, emissions of sulphur dioxide, nitrous oxides and other gases are controlled and reduced almost to zero from every piece of equipment in the plant, including the coke ovens, furnaces and boilers. Every piece of equipment that produces dust has its own electrostatic precipitator. Water is recycled indefinitely within the plant, with outside supplies needed only to replace water lost by evaporation. A computer-controlled 'energy centre' monitors the fluctuating demands of all components of the plant, and automatically allocates the optimum supply of coke, iron ore, electricity, steam, oxygen, natural gas, oil and water.

For Japan, such measures are no longer exceptional. The 'oil shokku' can never be forgotten. Energy conservation means not just maintaining the new prosperity, but perhaps survival itself. Even among the other industrial giants the OPEC initiative caused ominous tremors. The United States, once the world's greatest oil producer, felt particularly threatened. By the 1970s the Americans had developed a thirst for petroleum which could only be satisfied by a flood of imports. When the price of oil quadrupled, they began a drive to reduce energy wastage and find new wells.

The Americans also seized their own oil fields to wring every last barrel out of them. A systematic re-pumping is taking place all over the United States, but in no more unlikely location than the heart of Los Angeles. Here, concealed inside buildings which masquerade as warehouses or office blocks, more than one hundred wells are sucking oil from the fields which extend beneath the city, and which were first tapped at the end of the last century.

The sprawling suburbs of Los Angeles cover what was one of the richest oilfields in California. Huntington Beach, on the south side of the city, was once a black forest of oil derricks. Today it is a beach-side residential suburb, but there are still scores of wells yielding oil. Now, however, the derricks have gone. Instead, along the beach and boardwalk, in suburban gardens and back lots, beside busy highways, there are the nodding heads of small beam pumps, driven by electric motors. They are direct descendants of Thomas Newcomen's mighty Cornish pumps, and each lifts its few barrels a day.

In uptown Los Angeles the oil business is conducted on a much larger scale, and in a style appropriate to that casual but business-minded community. Between the soaring towers of Century City, the capital of the movie kingdom, there rises a tall, slender wooden obelisk, with no windows and only one small doorway, at the base. This alien spire in a forest of skyscrapers contains a full-size drilling rig, busy putting down another of the scores of new wells around the city that current world oil prices have made profitable.

And in that shrine to unreality, Hollywood itself, it should be no surprise to find that a huge office block on West Pico Boulevard encloses a minor oilfield. No fewer than forty wells plunge into the earth at diverging angles, and between them pump thousands of barrels of oil a day from the basement of this crowded city. Inside the huge building, which is simply an empty shell designed to contain noise and oil spray and escaping gas, 'roughnecks' and drillers work round the clock beneath tall derricks, pulling drill stems and adding pipe in a never-ceasing clangour as the wells go ever deeper.

So strict are the environmental controls on oil drilling in Los Angeles that few residents of the city, even those that pass the hidden wells on their way to work, are aware of what is going on. But drilling may become even more intensive. There is clearly a lot of oil still left under Los Angeles, as there is in old wells all over the world. The day must come, however, when the cost of these last-drop operations will rule them out.

The temporary fall in total oil consumption caused by the worldwide recession, and belated conservation programmes by the industrial nations, have postponed the day of reckoning when the rising curve of oil consumption finally cuts across the falling curve of production. Where then will the energy come from for the high-tech existence which has become the goal of nearly every human society? Other sources of energy will have to be found just to maintain present living standards in the West, let alone to meet the rising aspirations of the world at large.

Confronted with the prospect of an end to the 'free ride' provided by the planet's store of solar energy, handily packaged in the form of liquid hydrocarbons, one of man's first impulses was to reach directly for the Sun, as the ancient civilisations did. There is a certain nostalgia and comfort in that pervading warmth, and reassurance in its passive, non-threatening technology.

The application of solar energy in small units has not been difficult to achieve, and is proceeding worldwide. The French were the first to concentrate this benign and diffuse form of energy into a more intense and powerful medium, with their solar furnace near Odeillo, high in the Pyrenees. Sixty-three large flat mirrors reflect the Sun's rays on to a concave mirror, which in turn focuses the energy into a small chamber where temperatures of more than 3000°C are reached.

The first commercial production of electricity by solar energy on a continuing basis began in 1982 in the United States. Solar One in the Mojave Desert was built by the Southern Californian Edison Company public utility at a cost of more than one hundred million dollars, not only to develop operating techniques for solar power, but to provide electricity for its grid. More than one thousand highly-polished metal mirrors, arranged radially around a central tower and driven individually to follow the Sun, focus energy on to a boiler at the top of the tower. The heat raises steam, which drives generators at the foot of the tower.

If all the nations of the world are to be able to fulfil their needs in metals – whose production now consumes one third of the world's total energy output – alternatives to fossil fuels will soon have to be brought on stream. Solar energy is one potential source, although limited in its applications. It is not suited, for example, to smelting, which is a high energy consumer. *This is Solar One in the Mojave Desert in California, the first commercial solar generator of electricity.*

Unlike the older technologies for producing electricity, solar energy is eerily quiet. Around Solar One there is no sound but the wind, and no pollution of earth or atmosphere. This plant was expensive, but it works – at least on fine days – and there will no doubt be more like it. However, solar energy is limited in its applications by its low energy density (like the wind), compared to energy sources of high density, such as fossil fuels. It is also intermittent, and without some cheap method of storing electricity – which has yet to be invented – solar energy is unlikely to provide more than about five per cent of the world's energy needs before the twenty-first century, when fusion power, from reactions like those in the Sun itself, may become available.

The one new source of energy discovered this century which might sustain us into the next is at present under fearful reappraisal. The silent cooling towers of the Three Mile Island nuclear plant in Pennsylvania have thrown a long shadow across the most accessible path to an alternative source of power.

The unlocking of the great doors to the heart of matter may one day be reckoned the most formidable material achievement of the human intellect. The radioactive metals, in the energy that they offer, transcend in their possibilities all the other metals that mankind has ever put to work. The energy obtainable from a single drum of uranium can match the output of three hundred deep coal mines in Britain or fifty kilometres of strip-mining. A dramatic alternative beckons, therefore, to the accelerated burning of fossil fuels, and the growing apprehensions about their consequences, such as acid rain and the 'greenhouse effect' on the Earth's atmosphere.

But today, more than forty years after the first sustained chain reaction was achieved, only a few countries, particularly France and the USSR, have a serious commitment to nuclear energy. Other industrial nations such as Britain, Germany and Japan shuffle more obliquely forward, fearful and yet impelled by economic necessity. And the most advanced technological society the world has known hesitates and even backs away from the threshold of the new age. This reluctance should not perhaps seem surprising, for the attitude of millions around the world towards this new source of energy was clouded from the beginning by the way in which its awesome potential was first demonstrated.

At the northern end of the White Sands Missile Range in New Mexico, near a place called Alamagordo, there is a wide stretch of open desert beneath a huge blue dome of sky. Its flatness is relieved only by low sandhills and scattered patches of sagebrush. To the east there is a brooding range of mountains, whose silhouette is broken at intervals by the jagged craters of extinct volcanoes. A few cattle browse in the hazy distance, but there is no bird life. Apart from the soft sigh of the wind, a strange silence presses on the desert.

Standing in isolation in a clear space, a low, dark, pointed obelisk rises a few metres from the sand. It is built of blocks of black lava brought from the mountains, and it carries a bronze plaque with the words: 'Trinity Site. Where the World's First Nuclear Device was Exploded on July 16, 1945.'

The obelisk marks 'ground zero' at the base of the steel tower on which the device was exploded, in that ominous rehearsal for the destruction of Hiroshima and Nagasaki less than a month later. Here it was that man lit his last and most fiery furnace – the ultimate successor to all those other furnace fires which had glowed throughout history, and in which he had forged both his weapons and his wealth. Now, for the first time, he succeeded in reproducing on Earth, for an instant, something akin to the furnace of the Sun.

Under Dr Robert Oppenheimer, the greatest aggregate of scientific intellect ever brought together had worked against time at a heavily-guarded base at Los Alamos, in the Jemez Mountains north of

The black lava pillar in the desert of New Mexico, where man lit his last and most fiery furnace, symbolises the choice that faces the human race: between possible incineration of the planet, or the fruitful use of the awesome forces of the universe that it has taken us some three thousand years to understand. *The pillar marks 'ground zero', where the world's first atomic explosion took place on July 16, 1945, in that ominous rehearsal for the destruction of Hiroshima and Nagasaki less than a month later.*

Alamagordo, to construct – from purely theoretical calculations – the first atomic device. The Italian physicist Enrico Fermi had already demonstrated in Chicago in 1942 how to produce a sustained, controlled chain reaction. But the enormous potential of his work was overshadowed by the rumoured threat of a German atomic bomb. When Germany surrendered, the Manhattan Project still went ahead. It had generated its own urgent scientific momentum, and the Allied leaders were looking to this 'secret weapon' to precipitate Japan's collapse.

On 15 July the final assembly of the bomb's components was completed at a small adobe ranch house near the Alamagordo test site. Glowing gold in the afternoon sun, the steel ball containing the plutonium and the firing charge of high explosives was hoisted slowly to the top of the steel lattice tower. The electrical connections were made, and in dugouts and at observation points all around the desert Oppenheimer and his team waited for the longest night of their lives to pass.

Soon after darkness fell it began to rain, and thunder rolled and lightning crackled across the desert, as if to remind the huddled scientists of the verities. Just before midnight the arming party drove out to the tower, and one man climbed to the top and switched on the detonating circuit. In the early hours the rain began to ease, and Oppenheimer took the decision to go ahead, at 5.30 am instead of 4 am. Elsewhere the world slept on, or fought its war, and President Truman waited with his conscience in Potsdam.

As the last few minutes of the world as it was dragged past, the tension at the control centre grew almost unbearable. General T. F. Farrell wrote later: 'The scene inside the shelter was dramatic beyond words . . . It can be safely said that most everyone present was praying. Oppenheimer grew tenser as the seconds ticked off. He scarcely breathed. He held on to a post to steady himself.'

At 5.29:45 Mountain War Time the desert was lit, as Oppenheimer was to say later, 'brighter than a thousand Suns'. What followed was beyond the experience of the observers, but many have since tried to describe it. General Farrell was one: 'The effects could well be called unprecedented, magnificent, beautiful, stupendous, and terrifying. No man-made phenomenon of such tremendous power had ever occurred before. The lighting effects beggared description. The whole country was lighted by a searing light . . . it was golden, purple, violet, grey, and blue. It lighted every peak, crevasse and ridge of the nearby mountain range with a clarity and beauty that cannot be described but must be seen to be imagined. Seconds after the explosion came, first, the air blast pressing hard against the people, to be followed almost immediately by the strong, sustained awesome roar which warned of doomsday and made us feel we puny things were blasphemous to dare tamper with the forces heretofore reserved for the Almighty'.

William L. Laurance of the *New York Times*, the sole representative of the press, whose job it was to record the moment for history, wrote: 'It was like the grand finale of a mighty symphony of the elements, fascinating and terrifying, uplifting and crushing, ominous, devastating, full of great promise and great foreboding . . . It was as though the earth had opened and the skies split. One felt as though he had been privileged to witness the Birth of the World – to be present at the moment of Creation when the Lord said: "Let there be light".'

Oppenheimer was reminded of the ancient Hindu prophecy from the *Bhagavad Gita*: 'I am become Death, the destroyer of worlds.'[59]

When the scientists later came to the site they found the tower gone, vaporised. Where it had stood there was a huge shallow crater, eight hundred metres in diameter, lined with glass where the sand had been melted by the blast. Many years later, when the radioactivity had died down, the beaded glass saucer was bulldozed into rubble and buried. But the pale green, jade-like 'pearls of Alamagordo' can still be found in the sand around the black lava obelisk. Created by the nuclear fire in the New Mexico desert, they mark a great crossing, from the past into the future.

The atomic furnace lit by man that morning in July 1945 reversed the flow of all previous practice and experience. All the furnaces which had preceded this one consumed vast resources of energy – forests of trees, labyrinths of coal, underground oceans of oil – to produce relatively small quantities of matter, of copper and iron and aluminium. The nuclear furnace, fed with small resources of matter, yielded vast amounts of energy.

It might be said, then, that here at Alamagordo man gave up his wanderings in the deserts and the forests, and his habit of using whatever lay to hand, and turned to the ultimate source of energy. It had taken him some two thousand five hundred years, from the time the Greeks first began to speculate about the existence of the atom to that final Promethean exertion, lasting twenty-seven months, to demonstrate here that he could at last manipulate the fundamental stuff of the universe itself. With the nuclear furnace he could both consume and breed the new strategic metals of uranium and plutonium, which have taken the place of silver and gold in determining the conduct of relations between all-powerful nations.

But as Michael Charlton said in the final scene of our TV series, standing beside the black lava pillar in the New Mexico desert, 'Alamagordo could be the altar on which the whole of earth was set up as a burnt offering. But perhaps we should be more optimistic. In his monumental history, *The Decline and Fall*, Edward Gibbon reminded us that however much the opinion that the world is coming to an end deserves respect for its usefulness and its antiquity, it has not been found agreeable to experience'.

Nearly half a century on from Alamagordo we *are* still here, however precariously. And there is more than just survival to show for it. As millions around the world now watch the routine but still awesomely breathtaking launches of the space shuttle, it is clear that each new leap of technology is shadowed with risk. But that challenge has been indivisible from civilisation's historical progress, from the time when the perspectives of man were measured by the distance he could walk in a day, to his vistas now, out among the spheres.

Notes

Chapter One

1 p. 6 James Mellaart discusses these developing Neolithic societies in *Earliest Civilisations of the Near East*, published in 1978 by Thames and Hudson, London, as part of their *Library of Early Civilisations*. Mellaart describes his own major discovery in Anatolia in *Catal Huyuk*, published by Thames and Hudson in 1967.

2 p. 8 There is an interesting account of the role of obsidian in Neolithic life in an article in the March 1968 *Scientific American* entitled 'Obsidian and the Origins of Trade,' by J. F. Dixon, J. R. Cann, and Colin Renfrew.

3 p. 8 Cyril Stanley Smith, Professor Emeritus at the Massachusetts Institute of Technology, has written extensively on early metallurgy. This reference is to an article entitled 'Aesthetic Curiosity; the Root of Invention', published in the New York Times on 24 August, 1975.

4 p. 13 The assumption that copper was the first metal to be deliberately smelted is challenged by Noel H. Gale and Zofia Stos-Gale in their article 'Lead and Silver in the Ancient Aegean', in the June 1981 *Scientific American*. In this they argue that since lead is much easier to melt than copper, it may well have been the first metal to be smelted – perhaps in a camp fire. The Institute for Archaeo-Metallurgical Studies (IAMS) was formed in London in 1973 to support and co-ordinate international research into ancient mining and metallurgy. In 1978 IAMS became associated with the Institute of Archaeology at the University of London, and began courses in archaeo-metallurgy.

5 p. 18 The results of the various experiments are described by Beno Rothenberg, R. F. Tylecote, and P. J. Boydell in 'Chalcolithic Copper Smelting; Excavations and Experiments', available from IAMS.

6 p. 20 The full story of this remarkable discovery is described in *The Cave of the Treasure* by Pessah Bar-Adon, published in 1980 by the Israel Exploration Society. It contains many photographs and illustrations of the objects, with full scientific descriptions, and is available from the National Museum in Jerusalem.

7 p. 21 R. F. Tylecote's *A History of Metallurgy*, published in 1976 by the Metals Society in London, is the most comprehensive general account of man's use of metals, with valuable technical detail and an extensive reading list. It will be found a most useful reference for all chapters of this book. Professor Tylecote teaches at the Institute of Archaeology at the University of London.

8 p. 23 The first direct evidence of prehistoric copper smelting in the New World was discovered by two archaeologists, Izumi Shimada of Princeton University and Stephen Epstein of the University of Pennsylvania, and a geographer, Alan K. Craig of Florida Atlantic University. Their detailed account of the site at Batan Grande in northern Peru and of the smelting techniques employed there was published in *Science*, Vol. 216, p. 952. A digest of their paper appeared in the *New Scientist* of 17 June 1982, under the heading 'Copper smelting among the ancient Peruvians'.

Chapter Two

9 p. 27 The description of the extensive Egyptian smelting operations in the Sinai in the second millennium BC is based on personal observations by the author of the archaeological evidence there, in company with Beno Rothenberg, and on Rothenberg's book, *Timna*, published in 1972 by Thames and Hudson, London.

10 p. 33 A seminal study of this question is 'Evidence for the sources and use of tin during the Bronze Age of the Near East', by James D. Muhly and T. A. Wertime, published in *World Archaeology* in 1972. Theodore Wertime was one of the first academics to take a keen interest in the origins of metallurgy, and although his original view that the smelting process was too remarkable to have ever been 'invented' more than once by Neolithic man is no longer accepted, he is recognised as one of the major influences in the evolution of the discipline of archaeo-metallurgy. Muhly is a historian who has become one of the leading authorities on ancient metallurgy, particularly in the Near East.

11 p. 34 A description of the finding of the ingots is contained in the IAMS newsletter No. 1, 1980. Muhly also discusses their possible origin in 'New Evidence for Sources of and Trade in Bronze Age Tin', published in 1978 as a chapter in *The Search for Ancient Tin*, Smithsonian Institution, Washington DC.

12 p. 35 There is detailed discussion of the continuing mystery of ancient tin sources in Muhly's *Supplement to Copper and Tin; the Distribution of Mineral Resources and the Nature of the Metals Trade in the Bronze Age*, published in 1976 by Archon Books, Hamden, Connecticut.

13 p. 37 The initial discoveries of Bronze Age metal working in Thailand are described by Wilhelm G. Solheim II in the April 1972 *Scientific American*. In the article, 'An Earlier Agricultural Revolution', Solheim also puts forward the proposition that the domestication of food plants may have occurred in the South East Asia much earlier than in the Near East. Donn T. Bayard gives details of his excavation in 'Non Nok Tha: the 1968 Excavation; Procedure, Stratigraphy, and a Summary of the Evidence', 1971, *Studies in Prehistoric Anthropology No 4*, University of Otago, Dunedin, NZ.

14 p. 40 The many intriguing aspects of the Ban Chiang finds are described by the excavators of the site, Chester Gorman and Pisit Charoenwongsa, in the magazine *Expedition*, published by the University of Pennsylvania Museum in Philadelphia. Their well illustrated article, 'Ban Chiang: a Mosaic of Impressions from the First Two Years', appeared in Vol. 18, No. 4, 1976. In the same issue there is also an article 'The Techniques of the Early Thai metalsmith', by Tamara Stech Wheeler and Robert Maddin.

15 p. 42 Pisit Charoenwongsa's description of the important metal deposits in South East Asia is contained in his chapter, 'Early South East Asian Bronze in the Light of Excavation in Thailand', in the volume *Bronze Culture in Asia*, published in 1978 by the Asian Cultural Documentation Centre for UNESCO in Tehran. This book contains the reports and papers submitted at the UNESCO symposium on bronze culture held in Bangkok in July 1976.

16 p. 42 In 1982 the first public display of Ban Chiang artefacts outside Thailand was staged in the United States in a joint exhibition mounted by the Thai Department of

Fine Arts, the University of Pennsylvania Museum in Philadelphia, and the Smithsonian Institution Travelling Exhibition Service. As well as illustrations and descriptions of the artefacts, the catalogue contains Joyce White's illuminating essay 'The Ban Chiang Tradition: Artists and Innovators in Prehistoric Northeast Thailand.'

17 p. 43 In an article in the Ban Chiang exhibition catalogue, 'Ban Chiang in World Ethnological Perspective', Ward H. Goodenough, a professor and former chairman of the Department of Anthropology, University of Pennsylvania, and a specialist in the languages and cultures of Oceania, discusses the perplexing question of what he calls the 'truly amazing' distribution of the hundreds of Austronesian languages, and the light which may be thrown on this problem by the finds in Thailand. This subject is explored in greater detail by P. S. Bellwood, senior lecturer in prehistory at the Australian National University, in 'The Peopling of the Pacific' in the November 1980 *Scientific American*, and in his definitive work *Man's Conquest of the Pacific*, published by Collins, 1981.

18 p. 43 The hitherto unrecognised factors which may have enabled cave-dwelling hunter-gatherers in South East Asia to progress to the threshold of civilised life are discussed by Chester Gorman and Pisit Charoenwongsa in their paper 'From Domestication to Urbanisation: A South East Asian View of Chronology, Configuration and Change', delivered in 1978 at the symposium 'The origin of Agriculture and Technology: West or East?' in Aarhus, Denmark.

19 p. 44 There are many published works devoted to Shang bronze, but among the most comprehensive are *The Great Bronze Age of China*, 1980, Thames and Hudson, London, and *The Wonder of Chinese Bronzes*, 1980, Foreign Language Press, Beijing (Peking). Both these books cover not only the high point of Chinese bronze, the Shang period itself, but the subsequent periods – Western Zhou, Spring and Autumn, Warring States, and Western Han – in which, in the view of most critics, the inspiration and artistic discipline of the bronze craftsmen steadily declined.

20 p. 47 One of the most detailed and authoritative accounts of the development of Chinese bronze casting is *Metallurgical Remains of Ancient China*, by Noel Barnard and Sato Tamotsu, published in 1975 by Nichiosha in Tokyo. Barnard has published extensively on many aspects of ancient Chinese culture, including calligraphy, inscriptions, and metal technology. Professor Tamotsu is a member of the staff of the Department of Chinese Language and Literature at Ochanomizu University in Tokyo, and began his collaboration with Barnard while a research officer in the Department of Far Eastern History at the Australian National University in Canberra. The bulk of this massive work consists of lists (in Chinese) of all sites bearing metallurgical remains which have been excavated under properly controlled conditions; the lists are based on many hundreds of archaeological reports. This material is accompanied by a series of distribution maps (annotated in English) of all the metal artefacts, according to period, and a long essay on bronze casting and other metallurgical techniques.

Chapter Three

21 p. 55 The collection of more than three hundred copper artefacts from the Temple of Hathor in the Sinai presented the British Museum Research Laboratory with an unprecedented opportunity to analyse and study a large sample of Bronze Age metal work from a single site. The findings are described by Paul T. Craddock in British Museum Occasional Paper No. 20, entitled 'Scientific Studies in Early Mining and Extractive Metallurgy', published in 1980.

22 p.57 The result of Jane Waldbaum's exhaustive studies of possible factors involved in the beginning of the Iron Age were published in 1978 as 'From Bronze to Iron', Vol. 54 in the series *Studies in Mediterranean Archaeology*, Goteborg, Sweden.

23 p. 58 References to the Hittites are scattered through the standard histories of early civilisations, but a comprehensive account is to be found in the volume called *The Empire Builders*, by Jim Hicks, in the *Time-Life* series *The Emergence of Man*. This volume was published in 1974. A detailed description of the ruins of the Hittite capital is given in *Ancient Civilisations and Ruins of Turkey*, by Ekrem Akurgal, published in 1978 by the Haset Kitabevi (Hachette Library) in Istanbul, and available from the Museum of Anatolian Civilisations in Ankara.

24 p. 62 The essential discoveries of carburising, quenching and tempering of iron, which enabled the ancient blacksmith to transform iron from a metal inferior to bronze to one superior to it, are described by Robert Maddin, James D. Muhly and Tamara Stech Wheeler in 'How the Iron Age Began', in the October 1977 *Scientific American*. The role of the 'Sea Peoples' in this development, and the spread of the new iron technology, are discussed by the same three authors and K. R. Maxwell-Hyslop in a long paper, 'Iron at Taanach and Early Iron Metallurgy in the Eastern Mediterranean', published in the July 1981 *American Journal of Archaeology*.

25 p. 65 An excellent summary of the historical, cultural and technological impact of the Celts was published in the May 1977 *National Geographic*.

Chapter Four

26 p. 68 Little detailed description has been published so far on the finds from the Emperor Ch'in's 'terracotta army', but there is a quite comprehensive account in *The Great Bronze Age of China*', mentioned above (Note 19).

27 p. 73 The whole subject of Chinese metallurgy is covered by Sir Joseph Needham's monumental work *Science and Civilisation in China*, which is being published by Cambridge University Press in a number of volumes. Iron working is also treated at length by Needham in *The Development of Iron and Steel Technology in China*, published in 1958 by the Newcomen Society, London. Noel Barnard's *Metallurgical Remains of Ancient China* (see Note 20) also contains useful comments on Chinese iron-casting technology.

28 p. 77 A detailed account of the Han stack-casting techniques, with drawings and photographs, was published by the Chinese archaeologist Hua Jue-ming in the June 1983 *Scientific American*.

29 p. 78 An interesting series of radio discussions on Chinese social and technological history, including the possible reasons for the failure of the Chinese to industrialise at an early stage, was broadcast in 1981 by the Australian Broadcasting Commission. It has been published by the ABC under the title *Culture and Science in China*, and is available in ABC shops.

30 p. 79 *Aspects of Ancient Indian Technology* was published in 1979 by Banarsidass in Delhi. It is written by Dr H. C. Bhardwaj, who teaches archaeology and the history of science and technology at Banaras Hindu University. *Technology in Ancient India*, published in 1981, forms Volume 1 of a fifteen-volume history of science, technology and medicine in India, being written by Professor O. P. Jaggi and published by Atma Ram and Sons in Delhi and Lucknow. Professor Jaggi is Head of the Department of Clinical Research at the University of Delhi.

31 p. 82 Greek mining techniques, and the Roman technology which followed, are described by J. F. Healy in *Mining and Metallurgy in the Greek and Roman World*, published in 1978 by Thames and Hudson, London, as part of a series *Aspects of Greek and Roman Life*. Healy is an archaeologist, and Professor of Classics at Royal Holloway College, University of London.

32 p. 84 The long history of the Rio Tinto mine – without question the oldest mine in the world still in operation – has been meticulously compiled by David Avery in *Not On Queen Victoria's Birthday*, published in 1974 by Collins, London. (The odd title is a reference to the period of British operation of the mine, and the incomprehensibility to the Spanish workers of certain British ways – including the footnote in the timetable of the railway from the mine to the coast, which informed travellers that 'Trains do not run on Sundays or on Queen Victoria's birthday.') Of particular interest is the chapter entitled 'The Archaeology of Rio Tinto', which brings together what is generally known about the early workings. This knowledge has been greatly expanded as a result of the Huelva Archaeo-Metallurgical Project, which was set up in 1973 by the Universities of Madrid and Sevilla, and IAMS, London. These recent discoveries, and their interpretation, are given by Beno Rothenberg and Professor Antonio Blanco-Freijeiro of the University of Madrid in '*Ancient Mining and Metallurgy in South-West Spain*'. Published in 1981, this is the first volume in a new series, *Metals in History*, being produced by IAMS, London.

Chapter Five

33 p. 95 This explanation of the legend is related in *Connections*, by James Burke, published in 1978 by Macmillan London Limited. Burke's book is, incidentally, like Tylecote's *A History of Metallurgy*, a useful general reference for the present work, and particularly for the revival of technology in Europe after the Dark Ages, referred to later in this chapter.

34 p. 108 The Gutenberg Museum in Mainz contains a comprehensive assembly of material on the history of printing, and has published, in English, a most informative booklet on Gutenberg.

35 p. 108 The best and perhaps the only available English edition of Agricola's *De Re Metallica* is published by the Dover Press in New York, and it has an interesting history. In 1912 the original Latin text was translated by an American mining engineer, Herbert Hoover (later to become President of the United States) and his wife, Lou Henry Hoover. With the original 289 line drawings, it was published by the Hoovers in a limited edition, and also in *The Mining Magazine*, London, but eventually went out of print and became a rare collector's item. In 1950 the Dover Press produced, unaltered and unabridged, the 1912 edition, with all its illustrations, and with four facsimile pages of the original Latin text.

36 p. 116 A colourful account of the exploits of the Spanish conquest of the Aztecs and the Incas is contained in *The Conquistadors* by Hammond Innes, published in 1969 by Collins, London.

37 p. 117 In 1978 a superb exhibition called 'El Dorado; Colombian Gold' toured Australia. It was assembled from the vast collection of gold artefacts held by the Museo del Oro in Bogota, Colombia. The Australian catalogue contained a detailed account of Colombian metal-working techniques with gold, silver, copper and platinum, written by two members of the Museo staff, Clemencia Plazas de Nieto and Ana-Maria Falchetti de Saenz. The exhibition, augmented by loans from museums and private collections around the world, was then presented for six months at the Royal Academy in London. It was called 'The Gold of El Dorado', and formed perhaps the greatest display of pre-Colombian gold ever seen outside South America. In the London catalogue the text was expanded into a full technical treatise by Dr Warwick Bray, and this remains the best available guide to early South American metal working. A catalogue produced by the Oro del Peru Museum in Lima provides a shorter but still useful account of Inca and later Peruvian gold-working techniques.

Chapter Six

38 p. 124 Millions of words have been written about the gold rushes of the nineteenth century, beginning with the great Californian strikes in 1848, but a sharp, descriptive account of these hectic events is contained in *The Gold Rushes*, by Robin May, published in 1978 by Hippocrene books, New York. A more detailed history of all the momentous mineral discoveries in the United States is *A Pictorial History of American Mining*, by Howard and Lucille Sloan, published in 1970 by Crown, New York.

39 p. 130 Edward Hargraves' own account of his part in the first Australian gold rush, *Australia and its Goldfield*, was published in London in 1853, and is not generally available, although passages from it turn up in almost all later accounts. Geoffrey Blainey's assessment of those events, 'The Gold Rushes: The Year of Decision', was published in 1962 in *Historical Studies, Australia and New Zealand, Vol. 10, No. 38*, Melbourne University. Blainey's highly readable account of the later developments and consequences of the gold rushes elsewhere in Australia (which take up the remainder of this chapter) is called *The Rush That Never Ended*. It was first published by the Melbourne University Press in 1963, and has been reprinted at intervals ever since.

40 p. 136 G. Blainey, 'The Gold Rushes: The Year of Decision', 1962, *Historical Studies, Australia and New Zealand, Vol 10, No 38*, Melbourne University.

41 p. 137 *ibid.*

42 p. 138 G. Blainey, *The Rush That Never Ended*, 1963, Melbourne University Press.

Chapter Seven

43 p. 150 Uncountable volumes have been devoted to the Industrial Revolution, but this particular aspect of it has only recently received much attention. One useful article is 'An Early Energy Crisis and Its Consequences' by John Nef in the November 1977 *Scientific American*.

44 p. 155 L. T. C. Rolt was an authority on the great British engineers of the Industrial Revolution, and his books on Brunel, Telford, and the Stephensons are classics of nineteenth century biography. He was vice-president of the Newcomen Society, and president of the Association for Industrial Archaeology. John S. Allen is an engineer himself, and has specialised in research into the early stages of the development of the steam engine. Their book, *The Steam Engine of Thomas Newcomen*, was published in 1977 by Moorland Publishing Company, Buxton, Derbyshire, and Science History Publications, New York.

Chapter Eight

45 p. 172 Neil Cossons, *Industrial Archaeology*, 1975, David and Charles, London. Cossons is the director of the Ironbridge Gorge Museum Trust, whose chief concern is the preservation of the surviving industrial monuments in the Severn Valley, including Abraham Darby's original furnace, the Iron Bridge, and other relics of Coalbrookdale's period as the iron-making centre of the world. His book, besides listing and describing important archaeological sites in Britain, gives an excellent account of the development of wind and water power, coal and the steam engine, iron and steel, textiles and pottery, roads and bridges, canals, railways, and shipping.

46 p. 177 An entertaining, well illustrated account of Richard Trevithick's eventful life (including his adventures in South America, not of concern here), can be found in

Richard Trevithick by James Hodge, published in 1973 by Shire Publications, Aylesbury, Buckinghamshire.

47 p. 183 Of the many books on Brunel, perhaps the most absorbing is L. T. C. Rolt's *Isambard Kingdom Brunel*, which carries the subtitle: *Engineer, visionary and magnetic personality, he transformed the face of England.* It was first published by Longmans Green, London in 1957, but has since been re-published many times by Penguin Books.

Chapter Nine

48 p. 188 Among the more perceptive interpreters of the Industrial Revolution is Lord Asa Briggs, the eminent historian, now Provost of Worcester College, Oxford. *From Iron Bridge to Crystal Palace*, published in 1979 by Thames and Hudson, London, and richly illustrated from the archives of the Ironbridge Gorge Museum Trust and other sources, is an evocative portrayal of what Lord Briggs himself calls the 'impact and images of the Industrial Revolution'.

49 p. 188 This quotation is taken from an article in the twenty-fifth anniversary issue of the *New Scientist* on 9 July, 1981, which examined the current state of British science and technology. In his article, 'Britain's Engineering: Shadow of the Past', Simon Watt observed that the poor social status of engineers in Britain had its roots in the public's perception of their role as 'the instruments of the despised creators of wealth' in the nineteenth century. A recent book on this subject is M. J. Weiner's *English Culture and the Decline of the Industrial Spirit 1850–1980*, published by the Cambridge University Press.

50 p. 195 There is not a great deal in print on this intriguing episode in the story of man in North America, but the University of Michigan in Ann Arbor has published a collection of papers, entitled 'Lake Superior copper and the Indians – Miscellaneous Studies of Great Lakes History', edited by James B. Griffin. There is also a privately published booklet, *Copper Country History*, by Christine and Lauri Leskinen, residents of the Keweenaw Peninsula, which may be obtained through the museum at the Michigan Technological University in Houghton.

51 p. 198 The succession of remarkable mineral bonanzas discovered across the continent during this period in American history is well captured in the engravings and photographs in *A Pictorial History of American Mining* (see Note 38). Another excellent account, which also covers the history of mineral discoveries all over the world, is *Stones of Destiny* by John R. Poss, published in 1975 by the Michigan Technological University.

52 p. 201 *Hear That Lonesome Whistle Blow* by Dee Brown, published in 1977 by Pan Books, London, is one of the classics of American railroad history. Brown, who is the author of that moving account of the fate of the Indians, *Bury My Heart At Wounded Knee*, has caught the full flavour of the extraordinary saga of the Iron Horse.

53 p. 202 Brown, *ibid.*

Chapter Ten

54 p. 228 Professor Ronald King is responsible for the collection of Faraday's papers and scientific apparatus at the Royal Institution, and he has written a memorable account of the great scientist's life and achievements. *Michael Faraday of the Royal Institution*, a well-illustrated pamphlet published in 1973, is available from the Institution library.

55 p. 230 This account is from *The Beginning of the Incandescent Lamp and Lighting*

System, by Thomas Edison, published in 1976 by The Edison Institute, Dearborn, Michigan. It was written in 1926 at the request of Henry Ford, who wanted a record of this remarkable accomplishment in Edison's own words. Edison himself could claim no more than a few months of formal education in his life, and seldom wrote anything other than technical notes. He responded to his friend's request, however, and this account, although short, sets out lucidly the steps that he followed in those hectic months of creativity. It remains the only complete autobiographical work written by Edison.

56 p. 233 A good description of the evolution of this now universal system of open-cut mining is to be found in *Stones of Destiny* (see Note 51).

57 p. 234 Blainey, *The Rush That Never Ended* (see Note 39).

Chapter Eleven

58 p. 253 Among the swelling flood of publications on the subject of robots, one unusually interesting contribution is an article in the 12 February 1981 issue of *New Scientist,* entitled 'The Mechanisation of Mankind', by Peter Marsh. It traces the mechanisation of work through the ages, and points out that 'today's arguments about machines destroying jobs appear similar to many that have been voiced over the past seven hundred years at other times of advancing technology'.

59 p. 260 The description of that apocalyptic event is taken from an account recently prepared from previously published articles by the staff of the Los Alamos Scientific Laboratory, which is operated by the University of California for the United States Atomic Energy Commission. There are many books on the dawn of the atomic age, but one of the most unforgettable remains *Brighter Than a Thousand Suns,* by Robert Jungk, published in Germany in 1956 and in English in 1958 by Victor Gollancz, London. It is based on the vivid recollections of the many famous scientists, from ten nations, who were involved in the events leading up to July 16, 1945, and the consequences which have grown from it.

Index